Sustainable Manufacturing in Industry 4.0

Hamed Gholami · Georges Abdul-Nour ·
Safian Sharif · Dalia Streimikiene
Editors

Sustainable Manufacturing in Industry 4.0

Pathways and Practices

Springer

Editors
Hamed Gholami
Mines Saint-Etienne
Univ Clermont Auvergne, CNRS, UMR 6158 LIMOS, Institut Henri Fayol
Saint-Etienne, France

Safian Sharif
Department of Manufacturing and Industrial Engineering
Universiti Teknologi Malaysia (UTM)
Johor, Malaysia

Georges Abdul-Nour
Department of Industrial Engineering
University of Quebec in Trois-Rivieres
Trois-Rivieres, QC, Canada

Dalia Streimikiene
Lithuanian Energy Institute
Kaunas, Lithuania

ISBN 978-981-19-7217-1 ISBN 978-981-19-7218-8 (eBook)
https://doi.org/10.1007/978-981-19-7218-8

This Springer imprint is published by the registered company Springer Nature Singapore Pte Ltd.
The registered company address is: 152 Beach Road, #21-01/04 Gateway East, Singapore 189721, Singapore

Preface

Due to the sustainable development philosophy, popularized by the Brundtland report as "Our Common Future," the manufacturing sector has significantly evolved unprecedentedly over the years, unanimously accepting that being sustainable is more beneficial. Since the advent of the fourth industrial wave, a growing global research interest has emerged toward augmenting the economic, environmental, and societal values of Industry 4.0 through integrating its applicable technologies and intelligent techniques with a sustainable manufacturing paradigm, which seeks to develop more sustainable products—energy-efficient, eco-friendly, and socially-responsible—using sustainable processes and systems, i.e., those which produce minimal adverse environmental effects, conserve energy and natural resources, are harmless to people and viable to profit. To promote the research interest, this book offers an overview of the broad field of research on sustainable manufacturing in Industry 4.0. It includes the dissemination of original findings on pathways and practices of Industry 4.0 applied to sustainable manufacturing development, contributing new perspectives and roadmaps to those who are keen to realize the benefits of Industry 4.0 to transform the manufacturing sector into a more sustainable-based state.

This book features ten chapters. Chapters "A Review of Global Research Trends on Sustainable Manufacturing" and "An Analysis of the Literature on Industry 4.0 and the Major Technologies" shed light on how research on sustainable manufacturing and Industry 4.0 has evolved in recent years. Chapter "Smart Laser Welding: A Strategic Roadmap Toward Sustainable Manufacturing in Industry 4.0" investigates laser welding in the Industry 4.0 era and examines sustainable manufacturing challenges using an optimization-oriented perspective. The importance of additive manufacturing in sustainable manufacturing development is discussed in chapter "The Role of Additive Manufacturing in the Age of Sustainable Manufacturing 4.0". Chapter "The Impact of the Fourth Industrial Revolution on the Transitory Stage of the Automotive Industry" focuses on how Industry 4.0 paves the way for digitalization, smart manufacturing, and sustainability in the automotive industry and auto-guided vehicles. Policies on maintenance routines and the development of predictive maintenance tools as one of the pillars of sustainable and smart manufacturing are discussed in chapter "Advances in Smart Maintenance for Sustainable Manufacturing in Industry 4.0". A conceptual framework in the computer-aided inspection field with an extensive assessment is given in chapter "Smart Inspection; Conceptual Framework, Industrial Scenarios, and Sustainability Perspectives". Chapter "Sustainability Implications of Adopting Industry 4.0 at Different Scales in the Poultry Processing Industry" examines the converging points of sustainability and Industry 4.0 technologies in poultry processing plants. The latest smart technologies and systems in poultry processing, as well as the steps involved, were also discussed. Chapter "Horizontal Collaboration Business Model Towards a Sustainable I4.0 Value Creation" presents a CODAS-HTFLS-Mahalanobis approach to identify horizontal collaboration top variables grouped within the business model components, so as to create a value creation network towards sustainable manufacturing 4.0. Chapter "Assessment of Industry 4.0 Adoption for Sustainability in Small and Medium Enterprises: A Fermatean Approach" investigates how small and medium enterprises can adopt Industry 4.0 functions to achieve sustainability. A rigorous review to determine indicators of Industry 4.0 adoption in the context and a novel assessment method under Fermatean fuzzy sets were accordingly presented in this last chapter.

We sincerely acknowledge Springer for the given opportunity. We would also like to thank our colleagues Ramesh Nath Premnath and Silky Abhay Sinha for their support and guidance in completing the book. Finally, we would like to express our appreciation to all of the chapter contributors for their availability and valuable contributions.

Saint-Etienne, France — Hamed Gholami
Trois-Rivieres, Canada — Georges Abdul-Nour
Johor, Malaysia — Safian Sharif
Kaunas, Lithuania — Dalia Streimikiene

Contents

About the Editors

Dr. Hamed Gholami is an Assistant Professor of Industrial Engineering at the Ecole des Mines de Saint-Étienne and a Researcher at the Henri Fayol Institute, France. He received the M.Eng. degree in Industrial Engineering from the Universiti Teknologi Malaysia (UTM), in 2014, and the Ph.D. degree in Mechanical Engineering under the UTM International Doctoral Fellowship scholarship, in 2017. He was honored with the Best Ph.D. Student Award as well as the Alumni and Chancellor Awards at the UTM 59th Convocation Ceremony on Oct. 2017. Since January 2018, he had been a Postdoctoral Fellow with the Department of Manufacturing and Industrial Engineering, School of Mechanical Engineering, UTM. Dr. Gholami has also been acknowledged by the Malaysia Board of Technologists (MBOT) as a Professional Technologist in the field of Manufacturing and Industrial Technology. Given his knowledge and expertise in the field, he has served as a consultant and collaborator on granted projects, a member of editorial boards, a chief guest editor for special issues, and a reviewer in authoritative journals.

Dr. Georges Abdul-Nour is a Professor at the Department of Industrial engineering, Université du Québec à Trois-Rivières, Canada. He received his Bachelors and Master Degrees in IE from the Université de Moncton, N.B.' Canada. He completed his PhD at Texas Tech University; Lubbock Texas, U.S.A. Dr. Abdul-Nour authored and co-authored 10 books and books chapters, and more than 450-refereed papers in international journals and conferences proceeding, and technical reports. His research interest is in fields such as Intelligent Asset Management, Complexes Systems Resilience, Reliability and Maintenance, Intelligent Manufacturing, I4.0, Lean-Six Sigma, and JIT. He is a member of the Editorial Board of the International Journal of Production Research (IJPR), and on the Board of the International Society of Engineering Asset Management (ISEAM). Professor Abdul-Nour is a registered member of the Order of Engineers of Quebec (OIQ). He served as Chair of Engineering School and of Research Institute on Small and Medium Enterprises (SME), (INRPME) at UQTR. He is an Elected Member of the Academy of Industrial Engineering of Texas Tech. Dr. Abdul-Nour hold the HQ/NSERG Research Chair in Asset-Management and He is a Fellow of ISEAM.

Dr. Safian Sharif is a Professor at the Department of Materials, Manufacturing and Industrial Engineering, School of Mechanical Engineering, Universiti Teknologi Malaysia (UTM). He received his PhD from Coventry, United Kingdom in 2000. His main area of specialization is in machining of advanced materials such as titanium alloys, inconel, stainless steel and other exotic materials which include hardened steels, cobalt chromium molybdenum and composites. He is actively engaged in other research areas such as sustainable manufacturing, additive manufacturing, rapid tooling and injection moulding.

Dr. Dalia Streimikiene Ph.D. in Economics; Professor of Vilnius University and Leading Researcher at Lithuanian Energy Institute. She is working in the area of sustainable development since 1998 and is an internationally recognized expert in this field. She works on behalf of UNFCCC, European Commission, the United Nations Economic Commission for Europe, the United Nations Development Programme, World Energy Council, the IAEA etc. on sustainable energy, environmental and climate change mitigation policies. Dr. Streimikiene consults Lithuanian Ministry of Environment on Climate Change and participates in preparation of national reports to the UNFCCC Secretariat and the national greenhouse gas inventories. She is a member of State Commission for allocation of National Scientific Grants of Republic of Lithuania which is the most prestigious award by Lithuanian State for scientists. She represents expertise in Economics in this Commission. She has published over 200 articles in Web of Science (WoS) referred journals with high IF. She is a co-author of 15 monographs and presented many papers at many international conferences and forums around the world. She is Editor-in-Chief of Transformations in Business and Management journal indexed in WoS and SCOPUS databases. Her H index is 30 at Scopus.

A Review of Global Research Trends on Sustainable Manufacturing

Hamed Gholami, Falah Abu, Safian Sharif, Georges Abdul-Nour, and M. Affan Badar

1 Introduction

Despite having been used interchangeably in many cases, the terms 'Sustainable Development (SD)' and 'Sustainability' are inherently distinct—SD is the pathway to succeed in sustainability, that is the ideal dynamic state [1, 2]. A majority of the scientific community has been incorporating SD into the field of manufacturing, considering the growing global interest in the phenomenon as "Our Common Future [3]" drawn by the World Commission on Environment and Development, in 1987. The interest grew even larger following the revelation that our common future is intensively influenced by the manufacturing sector as revealed at the Earth Summit, Rio de Janeiro, Brazil in 1992 [4]. From that point on, the field has experienced numerous revolutions complying to the fact that being sustainable has greater benefits.

Being the core of all industrial economies, it was outlined that the manufacturing sector must be made sustainable with the aim of preserving the high standards of living already attained by industrialized societies and for enabling the sustainable achievement of the same standards of living by other developing societies. Thus,

H. Gholami (✉)
Mines Saint-Etienne, Univ Clermont Auvergne, CNRS, UMR 6158 LIMOS, Institut Henri Fayol, 42023, Saint-Etienne, France
e-mail: hamed.gholami@emse.fr

F. Abu · S. Sharif
Department of Manufacturing & Industrial Engineering, School of Mechanical Engineering, Universiti Teknologi Malaysia (UTM), 81310 Johor, Malaysia

G. Abdul-Nour
Department of Industrial Engineering, University of Quebec in Trois-Rivieres, Trois-Rivieres, QC G8Z 4M3, Canada

M. A. Badar
Department of Applied Engineering and Technology Management, Indiana State University Terre Haute, Terre Haute, IN 47809, USA

H. Gholami et al. (eds.), *Sustainable Manufacturing in Industry 4.0*,
https://doi.org/10.1007/978-981-19-7218-8_1

there is always a need for sustainable manufacturing development due to a number of prevailing issues such as the depletion of non-renewable resources, more stringent environmental and occupational safety/health regulations, and the growing penchant for environmentally-friendly products, among many others [5, 6, 7, 8]. Sustainable manufacturing entails the manufacturing of more sustainable products—energy-efficient, eco-friendly, and socially-responsible—by using sustainable processes and systems, i.e., those which produce minimal adverse environmental impacts, conserve energy and natural resources, are harmless to people, and are economically viable [9, 10, 11, 12]. However, according to [13], "there are many insufficient attempts, including a partially integral approach, almost all fall short because they largely deal with products and processes, but fail to stress the interconnectivity among the three integral elements involved in manufacturing (products, processes and systems), and show the basis for sustainable value creation an economic growth (p. 104)". This condition—a need for the development of sustainable products, processes, and systems—and the fact that this topic is dramatically receiving a great deal of attention from practitioners and researchers, thereby draws our fundamental question: how has research on sustainable manufacturing evolved in recent years?

To address the question, the current research carried out a Bibliometric or Scientometric analysis, which can expedite the review's process of research trends in the literature concerning the subject and subsequently give guidelines and directions for further investigations. This would contribute to providing up-to-date overview of the topic, including the possible implications for facilitating the complexities involved in the area of sustainable manufacturing. The methodological approach has been effectively employed since its inception in the early literature (i.e., [14, 15] which presented a description of Bibliometric research, up to its adoption in very recent studies [16, 17, 18, 19, 20, 21]. By using this method, the current study is primarily aimed at accomplishing the following objectives:

1. To present the past and present progress of the literature published on "sustainable manufacturing" and also its interchangeable term "sustainable production".
2. To characterise the most contributing countries to the understudied theme.
3. To recognise the core journals having a significant contribution to the subject.
4. To determine the highly contributing academic institutions to the under-researched topic.
5. To identify the prolific authors contributing considerably to developing the area.
6. To outline common terminology, research topics and in-depth insights.

Accordingly, this article is organized as follows: Sect. 1.2 clarifies the research methodological approach and the procedure of this study, Sect. 1.3 delivers findings of this overview and discusses the results according to the aforementioned objectives, and, finally, Sect. 1.4 provides the reader with a sense of closure on the topic.

2 Methods

Bibliometric analysis is a methodological approach which is applied to investigate the research trends in specific areas and outline the directions of such research through analysing the academic databases outputs [16, 22] according to co-occurrence, co-citation, co-author, co-word, and bibliographic coupling [17, 21]. Thus, this method has been carried out to examine global research trends in the area of sustainable manufacturing.

The data for this study was extracted from Scopus until May, 2021. However, the Scopus database is prominently regarded as the largest indexer of global research content, including titles from more than 5,000 publishers worldwide, e.g., Springer, ScienceDirect, Taylor & Francis, Emerald, Wiley, etc. [18, 23]. The bibliometric software and VOSviewer were accordingly used to statistically scrutinise the descriptive data including annually scientific production, most frequent keyword, and providing visualization for co-word analysis [19, 24].

2.1 Criteria for the Review

Similar to [25] and Guraja et al. [26], the documents considered for this review are limited to article, abstract report, book, book chapter, business article, conference paper, conference review, data paper, editorial, erratum, letter, multimedia, note, press release, report, retracted, review or short survey that were written only in English. We took into account all types of sources, book, book series, conference proceeding, journal, multi-volume reference works, newsletter, press release, report, and trade journal.

It is also mentioned that utilising the quotation marks ("") is essential to discover the exact phrases and to eschew lemmatization and synonym features of Scopus [20]. All the documents were filtered via article title, abstract and keywords to minimise duplication and undefined documents (without author's name). For data consistency, data from May 2021 onwards were not taken into account in this study.

2.2 Search Approaches for the Selection

The first search string used to analyse includes the keyword of "sustainable manufacturing (henceforth called as Sus-Man)", which resulted in a total of 1954 documents. The applied query was as follows: (TITLE-ABS ("sustainable manufacturing")) AND PUBYEAR < 2021 OR PUBDATETXT (("January 2021" OR "February 2021" OR "march 2021" OR "April 2021" OR "May 2021")) AND (EXCLUDE (PUBYEAR, 2022)) AND (LIMIT-TO (LANGUAGE, "English")). Then, the search string proceeds with the same course by replacing the term of "sustainable production" (henceforth called as Sus-Pro), resulting in a large number of 6392 documents from the Scopus database.

Next, the second part involves a combination of above search strings in an in-depth analysis, but it was limited to only journal and article types; however, the most common study designs for word search "sustainable production" OR "sustainable manufacturing" (henceforth called as Sus-Man/Pro) were journal articles (n = 4802, 58%). The query used was: (TITLE-ABS ("sustainable production" OR "sustainable manufacturing")) AND (LIMIT-TO (SRCTYPE, "j")) AND (LIMIT-TO (DOCTYPE, "ar")) AND PUBYEAR < 2021 OR PUBDATETXT (("January 2021" OR "February 2021" OR "march 2021" OR "April 2021" OR "May 2021")) AND (EXCLUDE (PUBYEAR, 2022)) AND (LIMIT-TO (LANGUAGE, "English")).

3 Results and Discussion

This section is completed through the procedure with the adopted methods according to the research objectives, as presented in Sect. 1.1. It discusses the detailed analyses and findings on each objective in an orderly manner in the ensuing segments.

3.1 Past and Present Progress of Research Interest

This segment presents the emerging trends in "sustainable production (Sus-Pro)" and "sustainable manufacturing (Sus-Man)" to provide a general outline of documents according to the author's keywords. As shown in Fig. 1, throughout the past forty-two years from 1979 to 2021, the research interest in Sus-Pro has acquired growing attention. An analysis of the temporal trend of the number of publications for Sus-Man was also performed. Interestingly, the keyword of Sus-Man is very common in Malaysia, which is ranked 11th among core contributing countries (outlined in Sect. 3.2). As a case in point, this term is commonly used by four prominent engineering/technology-based universities, which were found among the top fifteen contributors to the topic (explained in Sect. 3.4)—Universiti Teknikal Malaysia, Universiti Teknologi Malaysia, Universiti Tun Hussein Onn Malaysia, and Universiti Utara. However, the term 'sustainable manufacturing' was first reported after 13 years of publishing the oldest article, entitled "Markets for Alaskan oil", which had been aimed at developing the USA's economic, environmental, and national security goals [27].

The results show the research on this sustainable paradigm has considerably progressed, in particular, in the new millennium. A remarkable number of 857 documents were published on Sus-Pro in 2020 alone compared to 256 documents for Sus-Man (Fig. 1). The analyses indicate that the combination of publications on both Sus-Man and Sus-Pro (i.e., Sus-Man/Pro) were continuously increased every year

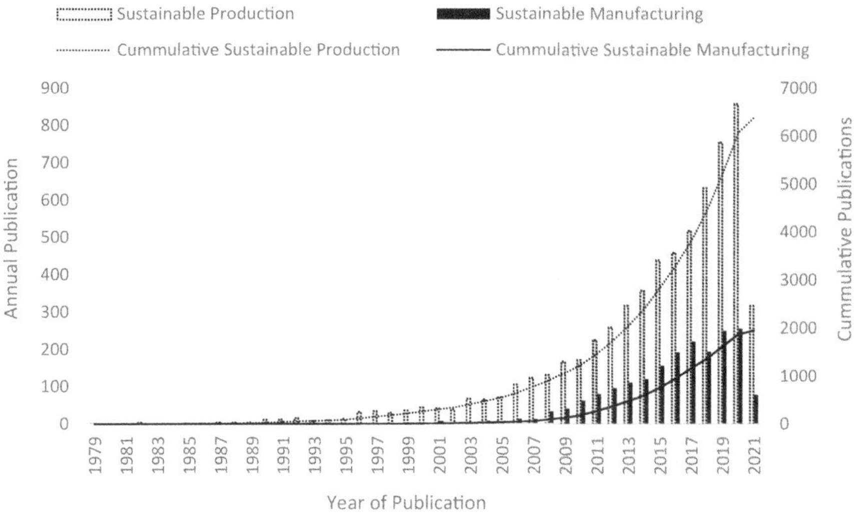

Fig. 1 Trend in publications over the years. Note: (i) Including all document type: article, abstract report, book, book chapter, business article, conference paper, conference review, data paper, editorial, erratum, letter, multimedia, note, press release, report, retracted, review paper, or short survey; (ii) Including all source type: book, book series, conference proceeding, journal, multi-volume reference works, newsletter, newspaper, press release, report, trade journal; (iii) 1954 documents on Sus-Man: (TITLE-ABS ("sustainable manufacturing")) AND PUBYEAR < 2021 OR PUBDATETXT (("January 2021" OR "February 2021" OR "march 2021" OR "April 2021" OR "may 2021")) AND (EXCLUDE (PUBYEAR, 2022)) AND (LIMIT-TO (LANGUAGE, "English")); (iv) 6392 documents on Sus-Pro: (TITLE-ABS ("sustainable production")) AND PUBYEAR < 2021 OR PUBDATETXT (("January 2021" OR "February 2021" OR "march 2021" OR "April 2021" OR "May 2021")) AND (EXCLUDE (PUBYEAR, 2022)) AND (LIMIT-TO (LANGUAGE, "English"))

since 2006, accordingly there was a dramatic growth in the cumulative total published documents hitherto. It is expected to continue to rise due to the unique intellectual contributor of Sus-Man/Pro to 'our common future'; however, it is unanimously accepted, after the Earth Summit [4], that being sustainable is more beneficial [17].

3.2 *Core Contributing Countries*

A total of 4802 journal articles published between 1979 and 2021on Sus-Man/Pro is dominated by developed and emerging countries. United States, China, India, the United Kingdom and Germany are the top five countries, respectively, as shown in Table 1. In terms of publication output, there is a huge gap between the top five countries identified. The United States tops the list with the publication of more than 800 research papers on the topic, followed by the United Kingdom and Germany at 4th and 5th places among developed countries, publishing less than 400 papers for the

Table 1 Country-wise growth of publications on sustainable manufacturing/production

Country	No. of articles[a]	National context
1. United States	817	Developed country
2. China	677	Emerging/developing country
3. India	470	Emerging/developing country
4. United Kingdom	341	Developed country
5. Germany	335	Developed country
6. Italy	291	Developed country
7. Brazil	231	Emerging/developing country
8. Netherlands	188	Developed country
9. Spain	181	Developed country
10. Australia	180	Developed country
11. Malaysia	172	Developing country

[a]Out of 4802 articles (document type) from journal (source type)

same time period. This is less than two times of the United States publication outputs, generally drawing attention to the environmentally harmful effects of manufacturing.

Nevertheless, it is interesting to look at the growth of publications on this area from the perspective of a developing country. Malaysia is the only developing country producing research outputs on Sus-Man/Pro after the top ten countries with 172 publications (Table 1). It is adjacent to Australia with less than 8 publications to be listed in the top ten based on its research outputs over the past years.

After conducting and merging the country profile and bibliometrics, the co-authorship analysis of countries provided 174 results. As such, we applied a threshold of a minimum of one document published per country and excluded any articles that co-authored more than 25 countries. A predetermined screening criterion was also used to screen and verify the list of countries. Unrelated terms such as "email", "university", etc. were discarded. Finally, a total of 139 countries were selected (Fig. 2).

As illustrated in Fig. 2, United States is the first core contributing country among others in all the parameters—total link strength (586), links (92), and documents (817, avg. pub. year: *ca.* 2014). The analyses also revealed that the most and recent co-author network is between United States and China. Based on the minimum link strength between countries, the first five countries, which had high collaborations with researchers from United States, are China, India, United Kingdom, Germany, and Australia. Meanwhile, the most co-author network for Malaysia was their regional neighbour, i.e., Indonesia. It is then followed by United Kingdom, Pakistan, and China.

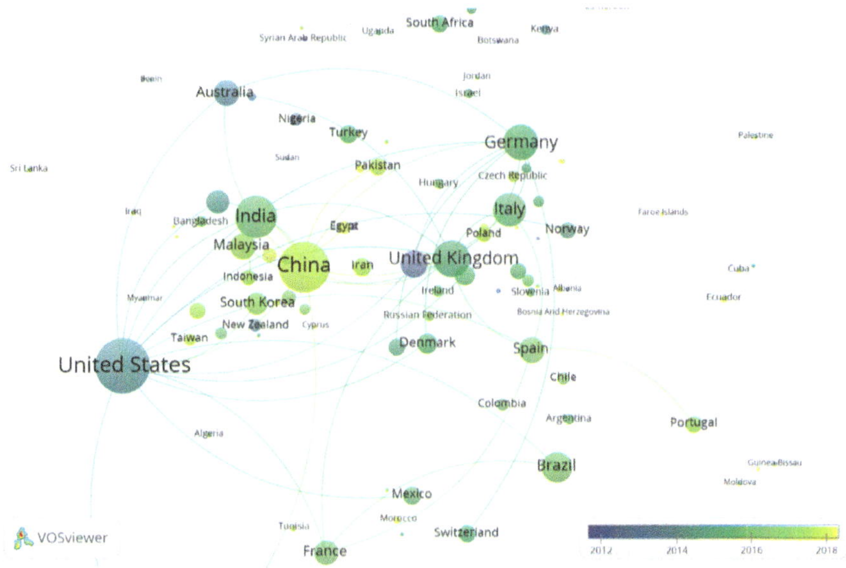

Fig. 2 Co-authorship network map of countries publishing on sustainable manufacturing/production

3.3 *Core Contributing Journals*

The findings indicate that 4802 articles are owned by 160 journals. The top 10 journals, with a share of 758 number of publications, are presented in Table 2. It is noticeable that six journals were from the United Kingdom and developing countries had none. The "*Journal of Cleaner Production*" published the maximum number of research articles on the understudied area, followed by the "*Sustainability*" and "*ACS Sustainable Chemistry and Engineering*".

Overall, 40% of the total research articles from the top 10 journals were published in the "*Journal of Cleaner Production*" (CiteScore 13.1), which remarkably includes the most cited article—[28]—among others (Table 2). The publication of research papers on this topic in these high-impact journals signifies the scientific community's growing interest and acknowledgement on the subject.

3.4 *Core Contributing Academic Institutions*

Research institutes from China has dominated the publications on research topic; Chinese Academy of Sciences (n = 89) and Ministry of Education China (n = 83). Starting with only 2 publications in 2012, Chinese Academy of Sciences had an incredible 89 published papers recently. With 172 publications in all, the Chinese

Table 2 Top 10 journals publishing research on sustainable manufacturing/production

Source title	Country	Publisher	Scopus Cite Score 2020	No. of articles[a]	Most cited article (times cited)
1. Journal of Cleaner Production	United Kingdom	Elsevier Ltd	13.1	305	[28] (503)
2. Sustainability	Switzerland	MDPI AG	3.9	155	[29] (131)
3. ACS Sustainable Chemistry and Engineering	United States	American Chemical Society	12.0	65	[30] (38)
4. Green Chemistry	United Kingdom	Royal Society of Chemistry	15.2	47	[31] (368)
5. International Journal of Advanced Manufacturing Technology	United Kingdom	Springer London	5.6	43	[32] (178)
6. International Journal of Production Research	United Kingdom	Taylor and Francis Ltd	10.8	33	[33] (84)
7. Bioresource Technology	United Kingdom	Elsevier Ltd	14.8	29	[34] (180)
8. Plos One	United States	Public Library of Science	5.3	28	[35] (218)
9. Procedia Manufacturing	Netherlands	Elsevier BV	13.1	28	[36] (65)
10. Biotechnology for Biofuels	United Kingdom	BioMed Central Ltd	9.9	25	[37] (93)

[a]Out of 4802 articles (document type) from journal (source type)

country is well ahead of other countries—see Table 3. About 55% of the publications from the top 10 research institutes come from emerging and developing countries, with more than 280 affiliated-published papers (Table 3). The continuously increasing publications from such countries is a clear sign that this field of research will only continue to grow in the near future.

Table 3 Top 10 research institutes working on sustainable manufacturing/production

Affiliation	Country	National context	No. of articles[a]
1. Chinese Academy of Sciences	China	Emerging/developing country	89
2. Ministry of Education China	China	Emerging/developing country	83
3. Wageningen University & Research	Netherlands	Developed country	78
4. USDA Agricultural Research Service	United States	Developed country	44
5. Universidade de Sao Paulo	Brazil	Emerging/developing country	44
6. United States Department of Agriculture	United States	Developed country	40
7. Empresa Brasileira de Pesquisa Agropecuária—Embrapa	Brazil	Emerging/developing country	35
8. Danmarks Tekniske Universitet	Denmark	Developed country	34
9. UNESP-Universidade Estadual Paulista	Brazil	Emerging/developing country	32
10. CNRS Centre National de la Recherche Scientifique	France	Developed country	32

[a]Out of 4802 articles (document type) from journal (source type)

3.5 *Core Contributing Authors*

Sekar Vinodh published a large number of articles on the topic with 18 research papers consistently every year since 2012, followed by two scientists, namely Fazleena Badurdeen and Norsiah Hami (Table 4). Interestingly, Norsiah Hami is the only scientist from developing country who was listed among the top three authors.

3.6 *Common Terminology, Research Topics and In-Depth Insights*

The investigation reveals that Sus-Man and Sus-Pro have been often applied interchangeably in the subject area of Engineering and Technology; however, there is also an inherent difference between them—'sustainable production' is a broader term that can be used in all subject areas. As shown in Fig. 3, 1034 out of 1954 documents were remarkably published on Sus-Man in the Engineering and Technology area

Table 4 Top 10 authors publishing on sustainable manufacturing/production

Author name	Institutions	Country	No. of articles[a]	Most cited article (times cited)
1. Vinodh, Sekar	National Institute of Technology Tiruchirappalli, India	India	18	[38] (64)
2. Badurdeen, Fazleena	University of Kentucky	United States	12	[8] (550)
3. Hami, Norsiah	Universiti Utara Malaysia	Malaysia	12	[39] (9)
4. Xu, Boqing	Tsinghua University	China	11	[40] (316)
5. Liang, Yu	Tsinghua University	China	10	[40] (316)
6. Gao, Liang	Huazhong University of Science and Technology	China	9	[41] (149)
7. Li, Lin	University of Illinois at Chicago	United States	9	[42] (147)
8. Pham, Duc Truong	University of Birmingham	United Kingdom	9	[43] (60)
9. Ocampo, Lanndon A	Cebu Technological University	Philippines	9	[44] (20)
10. Haapala, Karl R	Oregon State University	United States	8	[45] (73)

[a]Out of 4802 articles (document type) from journal (source type)

compared to 870 (out of 6392) documents for Sus-Pro, suggesting that 'sustainable manufacturing' is the most common term for such a subject area.

The co-occurrence analysis of keywords was accordingly performed for Sus-Man/Pro on a total of 4802 publications in 160 journals. A threshold of a minimum number of keywords occurrences equal to 5 was set. The analysis of Sus-Man/Pro resulted in 492 keywords out of a total of 13,466. Figure 4 displays the overlay visualization which is coloured differently based on the average publications' year. The overlay visualization ranges from white (old article) to dark purple (contemporary article). The dominant keywords based on total link strength were "sustainability" (612 total link strength), "sustainable manufacturing" (366), "sustainable development" (125) and "sustainable production" (118), respectively.

The analyses indicated that the links, total link strength, and occurrence for Sus-Man is ranked higher than Sus-Pro. Link is a connection or relation between two items (e.g., co-occurrence of keywords) while the total link strength is a weight

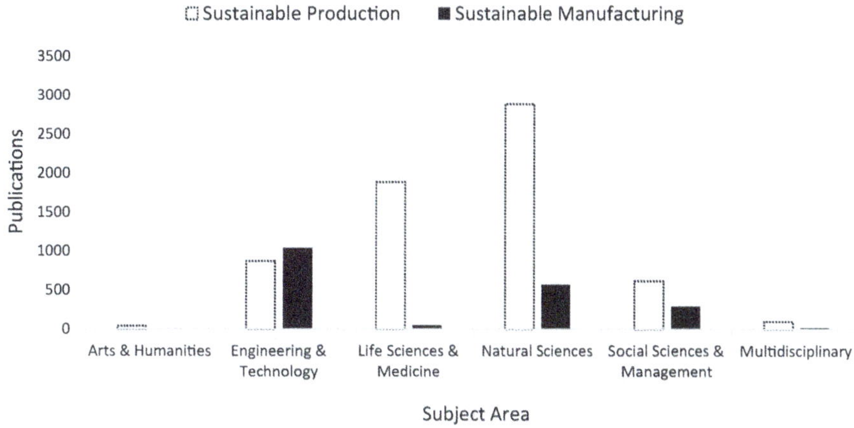

Fig. 3 Publications over the subject area

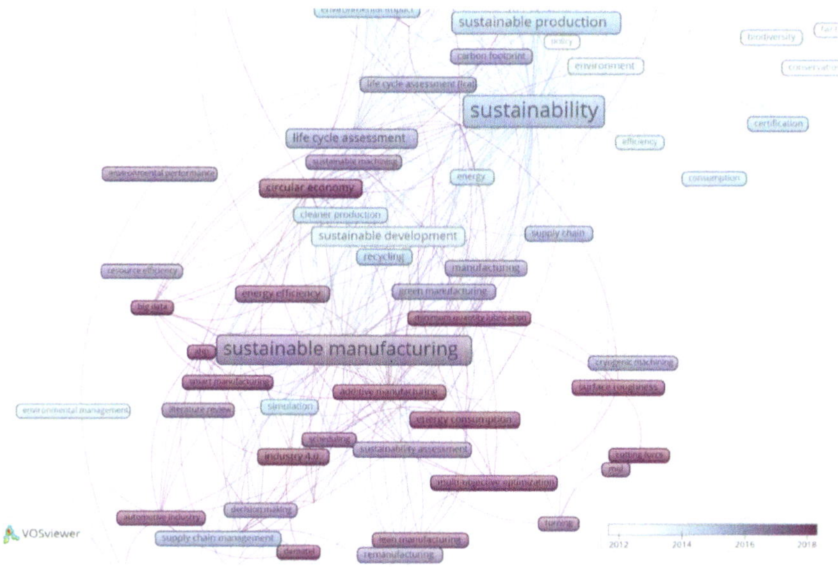

Fig. 4 Co-occurance network map of keywords from articles published on sustainable manufacturing/production

attribute to determine the cumulative strength of the links of an item with other items [24]. This could be explained by the fact that, the links and total link strength for Sus-Man are higher than Sus-Pro since it is a recent fascinated topic and thus more strongly connected to other emerging keywords such as ‘Circular Economy’ (total link strength: 72, avg. pub. year: 2018.9). The link between circular economy and sustainable manufacturing was well-argued by [13], who explained that circular economy can be operationalised in manufacturing through applying the 6Rs—Reduce, Reuse, Recycle, Recover, Redesign and Remanufacture. Other connected emerging keyword include 'Industry 4.0' (total link strength: 48, avg. pub. year: 2019.4).

Table 5 lists the 10 most influential articles on Sus-Man, which were ranked using Scopus in terms of the highest citation. The article by [8] received the highest citation count of 550, providing the all-inclusive overview of the concept by exemplifying the dry, near-dry and cryogenic machining. It is preceded by [46] with 436 citations, who concluded that the initiation of a new technology may modify the description of "what is sustainable". Noticeably, there is a paper in the list of the 10 most influential articles with 213 citations—[47]—which was very recently reported, among others, on a fundamental query about "can industry 4.0 revolutionise the environmentally-sustainable manufacturing wave?"

These global research trends depict a growing need for sustainable manufacturing development to sustainably address challenges and issues related to ecosystem destruction and numerous other unsustainable paradigms. There were many significant efforts as such; however, the development is generally traced by compartmentalising the manufacturing's integral elements—product, process, and system (Fig. 5). This may be due to sustainable manufacturing is a complex systems problem [13], and which it is being relied highly on the analytical approaches that make learning and development through the reductionism thinking and mechanism interpretation.

Figure 5 manifests a visual representation of elements, where the union is created by overlapping products (value design), processes (value creation), and systems (value recovery) based on the 6R methodology to fulfil the TBL requirements [1]. The colour gold was employed to denote sustainable development, thereby ensuring that SD is the Golden Pathway to manufacturing sustainability. Therefore, new technologies together with other critical success factors [46, 47] and mental models, on which the manufacturing encompasses interrelated elements, with interconnected processes, units, norms, values, behaviours, individuals and groups, which are influencing and being influenced by one another, are requested to sustainable manufacturing development [17].

4 Conclusion

This article presents an analysis of the research trends in sustainable manufacturing area using a bibliometric analysis in the Scopus database, which is prominently considered as the major indexer of global scientific content. The data for the study was extracted until May 2021 based on the descriptive data of publication outputs and resulted in retrieving a total of 4802 journal articles reported between 1979 and 2021. The bibliometric method contributed to provide the structures and development in the sustainable manufacturing area so that the scientific community could penetrate the existing hierarchy of the publication in the context. The analyses revealed that publication growth was swift; the published documents were continuously increased

Table 5 10 most influential articles on sustainable manufacturing.

Rank	Authors (Year)	Title	Source title	Times cited
1	Jayal et al. [8]	Sustainable manufacturing: modeling and optimization challenges at the product, process and system levels	CIRP Journal of Manufacturing Science and Technology	550
2	Garetti and Taisch [46]	Sustainable manufacturing: trends and research challenges	Production Planning and Control	436
3	Jovane et al. [48]	The incoming global technological and industrial revolution towards competitive sustainable manufacturing	CIRP Annals—Manufacturing Technology	282
4	Joung et al. [49]	Categorization of indicators for sustainable manufacturing	Ecological Indicators	279
5	Sarkis [50]	Manufacturing's role in corporate environmental sustainability—concerns for the new millennium	International Journal of Operations and Production Management	276
6	Rusinko [51]	Green manufacturing: an evaluation of environmentally-sustainable manufacturing practices and their impact on competitive outcomes	IEEE Transactions on Engineering Management	243
7	Yan and Li [52]	Multi-objective optimization of milling parameters—the trade-offs between energy, production rate and cutting quality	Journal of Cleaner Production	229
8	Ijomah et al. [53]	Development of design for remanufacturing guidelines to support sustainable manufacturing	Robotics and Computer-Integrated Manufacturing	214
9	Jabbour et al. [47]	When titans meet—can industry 4.0 revolutionise the environmentally-sustainable manufacturing wave? The role of critical success factors	Technological Forecasting and Social Change	213
10	Faulkner and Badurdeen [9]	Sustainable Value Stream Mapping (Sus-VSM): methodology to visualize and assess manufacturing sustainability performance	Journal of Cleaner Production	209

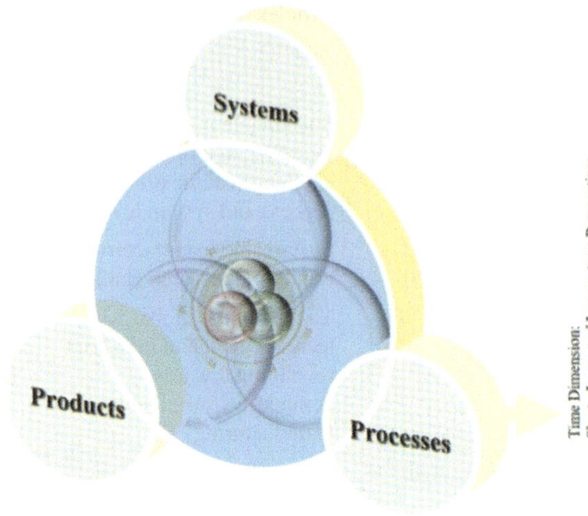

Fig. 5 Integral elements of sustainable manufacturing, from a general perspective to fully integrated perspective

every year since 2006. Core contributing countries, journals, academic institutions, and authors were also discovered. The United States and China are the countries in the top two, respectively, with an enormous number of publications and great collaboration networks. It may give an opportunity to investigators from other academic institutions and countries to widen their research collaborations. Furthermore, this study discussed some new areas considered for sustainable manufacturing which would be potential top topics for future research.

References

1. Gholami H, Saman MZM, Sharif S, Md Khudzari J, Zakuan N, Streimikiene D, Streimikis J (2020) A general framework for sustainability assessment of sheet metalworking processes. Sustainability 12(12):4957
2. Lozano R (2008) Envisioning sustainability three-dimensionally. J Clean Prod 16(17):1838–1846
3. WCED, 1987. Our Common Future, 3rd ed.; Oxford University Press: Oxford, UK, 24. UN.
4. Agenda 21 (1993) The United Nations Programme of action from Rio. Unite Nations, New York, NY, USA
5. Gholami H, Jamil N, Mat Saman MZ, Streimikiene D, Sharif S, Zakuan N (2021) The application of green lean Six Sigma. Bus Strateg Environ 30(4):1913–1931
6. Jamil N, Gholami H, Saman MZM, Streimikiene D, Sharif S, Zakuan N (2020) DMAIC-based approach to sustainable value stream mapping: towards a sustainable manufacturing system. Econ Res-Ekonomska Istraživanja 33(1):331–360
7. Jawahir I, Dillon Jr O (2007) Sustainable manufacturing processes: new challenges for developing predictive models and optimization techniques. In: Paper presented at the Proceedings of the first International Conference on Sustainable Manufacturing. Montreal, Canada
8. Jayal AD, Badurdeen F, Dillon OW Jr, Jawahir IS (2010) Sustainable manufacturing: Modeling and optimization challenges at the product, process and system levels. CIRP J Manuf Sci Technol 2(3):144–152

9. Faulkner W, Badurdeen F (2014) Sustainable value stream mapping (Sus-VSM): methodology to visualize and assess manufacturing sustainability performance. J Clean Prod 85:8–18
10. Gholami H, Jamil N, Zakuan N, Saman MZM, Sharif S, Awang SR, Sulaiman Z (2019) Social value stream mapping (Socio-VSM): Methodology to societal sustainability visualization and assessment in the manufacturing system. IEEE Access 7:131638–131648
11. US DOC (2010) The International Trade Administration and the U.S. Department of Commerce's Definition for Sustainable Manufacturing. Retrieved from http://www.trade.gov/competitiveness/sustainablemanufacturing/how_doc_defines_SM.as
12. Verma N, Sharma V, Badar MA (2021) Entropy-Based Lean, Energy and Six Sigma Approach to Achieve Sustainability in Manufacturing System. Arab J Sci Eng 46(8):8105–8117. https://doi.org/10.1007/s13369-021-05826-x
13. Jawahir I, Bradley R (2016) Technological elements of circular economy and the principles of 6R-based closed-loop material flow in sustainable manufacturing. Procedia CIRP. 40(1):103–108
14. Broadus R (1987) Toward a definition of bibliometrics. Scientometrics 12(5–6):373–379
15. Pritchard J (1969) Statistical bibliography or bibliometrics? J Doc 25(4):348–349
16. Abu F, Gholami H, Saman MZM, Zakuan N, Sharif S, Streimikiene D (2021) Pathways of lean manufacturing in wood and furniture industries: a bibliometric and systematic review. Eur J Wood Wood Prod, 1–20
17. Gholami H, Abu F, Lee JKY, Karganroudi SS, Sharif S (2021) Sustainable manufacturing 4.0—pathways and practices. Sustainability 13(24):13956
18. Letchumanan LT, Gholami H, Yusof NM, Ngadiman NHAB, Salameh AA, Štreimikienė D, Cavallaro F (2022). Analyzing the factors enabling green lean six sigma implementation in the industry 4.0 Era. Sustainability 14(6):3450
19. Lim ZS, Wong RR, Wong CY, Zulkharnain A, Shaharuddin NA, Ahmad SA (2021) Bibliometric analysis of research on diesel pollution in antarctica and a review on remediation techniques. Applied Sciences (Switzerland) 11(3):1–13
20. Ubando AT, Africa ADM, Maniquiz-Redillas MC, Culaba AB, Chen WH, Chang JS (2021) Microalgal biosorption of heavy metals: a comprehensive bibliometric review. J Hazard Mater, 402
21. Zahri KNM, Zulkharnain A, Sabri S, Gomez-fuentes C, Ahmad SA (2021) Research trends of biodegradation of cooking oil in Antarctica from 2001 to 2021: a bibliometric analysis based on the scopus database. Int J Environ Res Public Health 18(4):1–15
22. Khudzari JM, Kurian J, Tartakovsky B, Raghavan GV (2018) Bibliometric analysis of global research trends on microbial fuel cells using Scopus database. Biochem Eng J 136:51–60
23. Lee JKY, Gholami H, Saman MZM, Ngadiman NHAB, Zakuan N, Mahmood S, Omain SZ (2021) Sustainability-oriented application of value stream mapping: a review and classification. IEEE Access 9:68414–68434
24. Pauna VH, Picone F, Guyader GL, Buonocore E, Franzese PP (2018) The scientific research on ecosystem services: a bibliometric analysis. Ecol Quest 29(3):53–62
25. Al-Odeh MA, Smallwood J, Badar MA (2021) A framework for implementing sustainable supply chain management. Int J Adv Oper Manag 13(3):212–233
26. Guraja PK, Badar MA, Moayed FA, Kluse CJ (2022) Systematic literature review of the impact of state budget cuts on public higher education institutions in the U.S., Proceedings of the 12th international conference on industrial engineering and operations management, Istanbul, Turkey, March 7–10, 2022.
27. Hwa D, Klema E, Mancke R (1979) Markets for Alaskan oil. Energy Policy 7(1):23–28
28. Papargyropoulou E, Lozano R, Steinberger JK, Wright N, bin Ujang Z (2014) The food waste hierarchy as a framework for the management of food surplus and food waste. J Clean Prod 76:106–115
29. Ebert AW (2014) Potential of underutilized traditional vegetables and legume crops to contribute to food and nutritional security, income and more sustainable production systems. Sustainability 6(1):319–335

30. Bian H, Luo J, Wang R, Zhou X, Ni S, Shi R, Fang G, Dai H (2019) Recyclable and reusable maleic acid for efficient production of cellulose nanofibrils with stable performance. ACS Sustain Chem Eng 7(24):20022–20031
31. Katryniok B, Kimura H, Skrzyńska E, Girardon JS, Fongarland P, Capron M, Ducoulombier R, Mimura N, Paul S, Dumeignil F (2011) Selective catalytic oxidation of glycerol: perspectives for high value chemicals. Green Chem 13(8):1960–1979
32. Dubey R, Gunasekaran A, Childe SJ, Wamba SF, Papadopoulos T (2016) The impact of big data on world-class sustainable manufacturing. Int J Adv Manuf Technol 84(1–4):631–645
33. Dubey R, Gunasekaran A, Chakrabarty A (2015) World-class sustainable manufacturing: framework and a performance measurement system. Int J Prod Res 53(17):5207–5223
34. Gupta R, Sharma KK, Kuhad RC (2009) Separate hydrolysis and fermentation (SHF) of Prosopis juliflora, a woody substrate, for the production of cellulosic ethanol by Saccharomyces cerevisiae and Pichia stipitis-NCIM 3498. Biores Technol 100(3):1214–1220
35. McDonald RI, Fargione J, Kiesecker J, Miller WM, Powell J (2009) Energy sprawl or energy efficiency: climate policy impacts on natural habitat for the United States of America. PLoS ONE 4(8):e6802
36. Waibel MW, Steenkamp LP, Moloko N, Oosthuizen GA (2017) Investigating the effects of smart production systems on sustainability elements. Procedia Manuf 8:731–737
37. Angermayr SA, Van der Woude AD, Correddu D, Vreugdenhil A, Verrone V, Hellingwerf KJ (2014) Exploring metabolic engineering design principles for the photosynthetic production of lactic acid by Synechocystis sp. PCC6803. Biotechnol Biofuels 7(1):1–15
38. Thirupathi RM, Vinodh S (2016) Application of interpretive structural modelling and structural equation modelling for analysis of sustainable manufacturing factors in Indian automotive component sector. Int J Prod Res 54(22):6661–6682
39. Hami N, Muhamad MR, Ebrahim Z (2016) The impact of sustainable manufacturing practices on sustainability. Jurnal Teknologi 78(1):139–152
40. Chai SH, Wang HP, Liang Y, Xu BQ (2007) Sustainable production of acrolein: investigation of solid acid–base catalysts for gas-phase dehydration of glycerol. Green Chem 9(10):1130–1136
41. Lu C, Gao L, Li X, Pan Q, Wang Q (2017) Energy-efficient permutation flow shop scheduling problem using a hybrid multi-objective backtracking search algorithm. J Clean Prod 144:228–238
42. Wang Y, Li L (2013) Time-of-use based electricity demand response for sustainable manufacturing systems. Energy 63:233–244
43. Pham DT, Thomas AJ (2011) Fit manufacturing: a framework for sustainability. J Manuf Technol Manag 23(1):103–123
44. Ocampo LA (2019) Applying fuzzy AHP–TOPSIS technique in identifying the content strategy of sustainable manufacturing for food production. Environ Dev Sustain 21(5):2225–2251
45. Zhang H, Haapala KR (2015) Integrating sustainable manufacturing assessment into decision making for a production work cell. J Clean Prod 105:52–63
46. Garetti M, Taisch M (2012) Sustainable manufacturing: trends and research challenges. Prod Plan Control 23(2–3):83–104
47. Jabbour ABLS, Jabbour CJC, Foropon C, Godinho Filho M (2018) When titans meet–Can industry 4.0 revolutionise the environmentally-sustainable manufacturing wave? The role of critical success factors. Technol Forecast Soc Chang 132:18–25
48. Jovane F, Yoshikawa H, Alting L, Boer CR, Westkamper E, Williams D, Tseng M, Seliger G, Paci AM (2008) The incoming global technological and industrial revolution towards competitive sustainable manufacturing. CIRP Ann 57(2):641–659
49. Joung CB, Carrell J, Sarkar P, Feng SC (2013) Categorization of indicators for sustainable manufacturing. Ecol Ind 24:148–157
50. Sarkis J (2001) Manufacturing's role in corporate environmental sustainability—concerns for the new millennium. Int J Oper Prod Manag 21(5/6):666–686
51. Rusinko C (2007) Green manufacturing: an evaluation of environmentally sustainable manufacturing practices and their impact on competitive outcomes. IEEE Trans Eng Manage 54(3):445–454

52. Yan J, Li L (2013) Multi-objective optimization of milling parameters—the trade-offs between energy, production rate and cutting quality. J Clean Prod 52:462–471
53. Ijomah WL, McMahon CA, Hammond GP, Newman ST (2007) Development of design for remanufacturing guidelines to support sustainable manufacturing. Robot Comput-Integr Manuf 23(6):712–719

An Analysis of the Literature on Industry 4.0 and the Major Technologies

Hamed Gholami, Falah Abu, Jocelyn Ke Yin Lee, and Georges Abdul-Nour

1 Introduction

The value creation in industrialised nations is currently being shaped by Industry 4.0 (I4.0), which is a new wave of industrialisation seeking to develop Cyber-Physical Systems (CPS) through an amalgamation between manufacturing operations systems and information and communication technologies (ICT), particularly the Internet of Things (IoT) [1, 2]. Following the initial three industrial revolutions, as shown in Fig. 1, I4.0 possesses great potential to significantly improve industrial productivity via profound changes in the interrelatedness of the systems. A CPS integrates information technology with the operational technologies of the physical system. Meanwhile, the IoT refers to an ecosystem of technologies that monitors the status of physical objects, captures significant data, and communicates that data to software applications through established networks. In I4.0, the CPS elements are essentially linked by IoT [3]. Other technologies that are mainly deployed in I4.0 include Big Data Analytics, Industrial IoT, Simulation/Optimization, Additive Manufacturing, Horizontal/Vertical System Integration, Virtual/Augmented Reality, Autonomous Robots, Cloud, and Cybersecurity [4–7].

Following this concept, a number of countries have developed their own programs to accelerate the adoption and advance of I4.0 technologies. The birthplace of the

H. Gholami (✉)
Mines Saint-Etienne, Univ Clermont Auvergne, CNRS, UMR 6158 LIMOS, Institut Henri Fayol, 42023, Saint-Etienne, France
e-mail: hamed.gholami@emse.fr

F. Abu · J. K. Y. Lee
Department of Manufacturing & Industrial Engineering, School of Mechanical Engineering, Universiti Teknologi Malaysia (UTM), 81310 Johor, Malaysia

G. Abdul-Nour
Department of Industrial Engineering, University of Quebec in Trois-Rivieres, Trois-Rivieres, QC G8Z 4M3, Canada

H. Gholami et al. (eds.), *Sustainable Manufacturing in Industry 4.0*,
https://doi.org/10.1007/978-981-19-7218-8_2

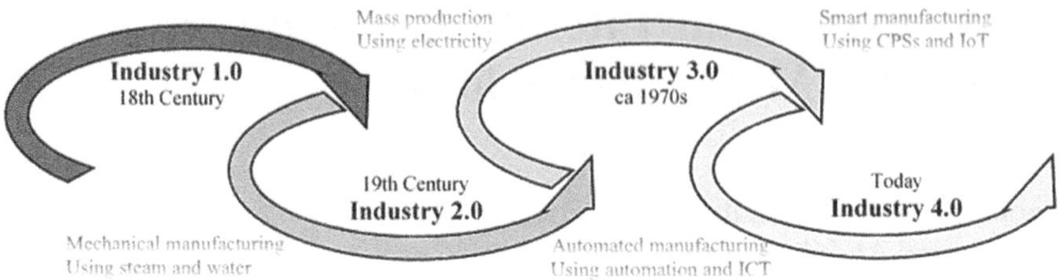

Fig. 1 Evolution of industrial waves

concept, i.e., Germany, had developed a program named "High-Tech Strategy 2020". The United States developed its "Advanced Manufacturing Partnership", France with "La Nouvelle France Industrielle", China with "Made in China 2025", and Brazil with "Towards Industry 4.0" (Rumo à Indústria 4.0). Such local programs, whether in developed or emerging countries, have the objective of disseminating the concepts and technologies of I4.0 to local businesses [1]. This implies that such countries have already conceived I4.0 concepts and technologies and subsequently matured with regard to the two concepts of Industry 3.0—automation and ICT usage—which are now being incorporated into I4.0 [8]. As I4.0 mainly entails the diffusion and adoption of technology, developing countries may encounter challenges in the form of a sluggish diffusion-adoption process as it typically flows from developed countries [1, 9, 10]. To demonstrate how this process is scientifically progressing, this study employed Bibliometric analysis, which is widely used for examining and analysing massive volumes of scientific data and also for investigating evolutions and possible areas of research in a specific field.

With the substantially growing academic and industry interests in I4.0, hereby this paper by using Bibliometric analysis aims to: (i) analyse journal articles on I4.0 for temporal distribution patterns, (ii) demonstrate the contributions of leading countries, prolific authors, and top educational establishments, (iii) underline common terminologies, and (iv) outline the most major technologies distributed throughout I4.0 publications. Implications wise, this work may facilitate industrial policy-/decision-makers, practitioners, and research experts in understanding how the emerging concepts and technologies of this new industrial wave have been scientifically progressed hitherto.

2 Methods

Bibliometric is a systematic analysis approach that uses statistics and quantitative analysis to explore trends in global research in a particular domain [11, 12]. Bibliometric analysis uses keywords from the titles of the documents, keywords, and summaries to find the connections between terms [13]. Such analysis allows researchers to discover the knowledge structure, challenges, and future orientations of the topic [14, 15].

2.1 Search Strategy and Data Collection

The dataset for the study was collected on July 26th, 2021 using Scopus. The central theme was journal articles comprising "Industry 4.0" in the title, abstract, and keywords. According to [16], the limitation of bibliometric review is that it may not take into account all of the most important characteristics of the text, and there are co-existing words that are closely related. Thus, this research has chosen the thematic search to follow the keywords related to I4.0 at once, which are "IR 4.0" and "fourth industrial revolution".

Firstly, we used the following combinations of queries: (TITLE-ABS ("industry 4.0" OR "IR 4.0" OR "fourth industrial revolution")) AND PUBYEAR <2021 OR PUBDATETXT (("january 2021" OR "february 2021" OR "march 2021" OR "april 2021" OR "may 2021")) AND (EXCLUDE (PUBYEAR, 2022)) AND (LIMIT-TO (LANGUAGE, "English")). This query string resulted in 11,755 publications with the oldest publication dating to 2003.

Next, additional terms were inserted into the query string to make sure that we did not include any review articles in the analysis; the documents were limited to articles (LIMIT-TO (DOCTYPE, "ar")) and journals (LIMIT-TO (SRCTYPE, "j")). Results revealed that the papers involved in this research were available in six distinct languages. English with 3993 documents (98.35% of overall publications) was the most prevalent language among German (57; 1.40%), Italian (6; 0.15%), Spanish (2; 0.05%), Czech and Portuguese (1; 0.02%), respectively. However, once a publisher submits a paper in a non-English language to Scopus for indexing, the publication should also contain an English title and abstract. Following that, any irrelevant articles to the keywords IR 4.0 were disregarded through reading titles and abstracts. As a result, 3988 articles were extracted for further analysis, as shown in Fig. 2. Nonetheless, if the keywords or topics did not involve any of the investigated phrases, it is possible that this search missed relevant articles out of the sample [16, 17].

2.2 Bibliometric Maps and Analysis

We utilized the Scopus database, VOSviewer version 1.6.16, and Publish/Perish (PoP) software in the research analysis processes. Firstly, the entire record metadata and cited references were obtained from the Scopus database. They were then transferred into VOSviewer to be further cleaned up as well as into PoP software for additional investigation. In this regard, we used the main technical terms applied by VOSviewer—Links, Citation links, Co-citations links, Co-authorship links, Co-occurrence links, Network, Weight attribute (Total link strength), Custom weight

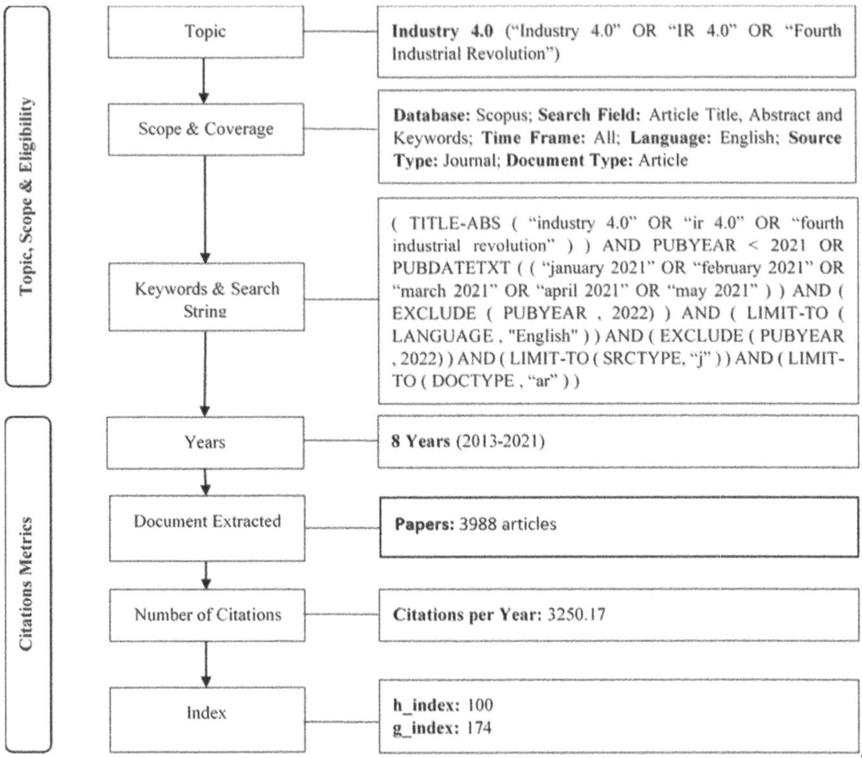

Fig. 2 Flow diagram of the search strategy

attributes (Documents, Occurrences, and Total citations), Density and overlay visualisation maps, Average citations per publication, h-index, and g-index—in pursuit of [12, 18–21].

The search output patterns were accordingly compared between the central theme (keyword co-occurrences) and the sub-theme (total publication). As extracted keywords from the Scopus database are not similar to the VOSviewer software, a thesaurus file must be created to export selected keywords and subsequently to group all the similar keywords involving the same terms. For instance, the total publications in the Scopus database for the keyword "Internet of Things" is 432 documents (10.82% of overall publications), followed by "Internet of Things (IOT)" 116 (2.91%), "IoT" 104 (2.60%), "Internet of Things (IoT)" 68 (1.70%), and "Internet Of Thing (IOT)" 66 (1.65%). Therefore, if "Internet of Things" is the keyword, hence keywords occurrences in VOSviewer software for "Internet of Things", "Internet of Things (IOT)", "IoT", "Internet of Things (IoT)", and "Internet of Thing (IOT)" were all counted as one term. Similarly, the following keywords were considered for the analysis, which is discussed in the ensuing section: Big Data

Analytics, Industrial IoT, Simulation/Optimization, Additive Manufacturing, Horizontal/Vertical System Integration, Virtual/Augmented Reality, Autonomous Robots, Cloud, and Cybersecurity.

3 Results and Discussion

This section follows the procedure based on the adopted methods for achieving the research objectives, as delineated in Sect. 1, accordingly presenting the analyses and findings in the subsequent segments.

3.1 Historical and Cumulative Growth Trend

Based on our formulated search string, as discussed in the previous section (Fig. 2), a total of 3988 journal articles written in English have been published over a span of eight years. Table 1 outlines the historical evolution of the major descriptors of scientific outputs from 2013 to 2021. In 2013, Ziesing and Hochstein published a paper, entitled "*Engineering tools as a basis for industry 4.0*", which is recorded by Scopus as the oldest publication in the I4.0 context. [22] describe the case study of ThyssenKrupp company that is working with complex computer-aided tools and augmented in engineering activities. The growth in the number of documents recorded in the next three years was not promising (Table 1), rising from the aforementioned article published in 2013 to 92 articles in 2016.

Table 1 Descriptive data of I4.0 publication outputs over the years

Year	TP	%	NCP	TC	C/P	C/CP	*h*	*g*
2021	446	0.11	235	957	2.15	4.07	14	19
2020	1633	0.41	1154	8962	5.49	7.77	37	53
2019	968	0.24	803	11,758	12.15	14.64	45	77
2018	519	0.13	456	15,621	30.10	34.26	62	107
2017	274	0.07	243	11,816	43.12	48.63	56	103
2016	92	0.02	75	4095	44.51	54.60	26	63
2015	41	0.01	33	3456	84.29	104.73	15	33
2014	14	0.00	10	1559	111.36	155.90	6	10
2013	1	0.00	0	0	0.00	0.00	0	0

Notes TP = total number of publications; NCP = number of cited publications; TC = total citations; C/P = average citations per publication; C/CP = average citations per cited publication; h = h-index; and g = g-index

Since this year, annual publications have continuously grown, leading to a rapid rise in cumulative overall publications. Significantly, publications on the topic have dramatically increased to 1633 articles in 2020. This trend shows an increasing interest in the subject as well as an expectation that the annual publications will keep growing due to the diffusion-adoption process which typically flows from the contributing countries.

3.2 Contributing Countries on the Topic of Concern

Table 2 displays the top ten most productive countries that have considerably contributed to the global expansion of I4.0 research activities. Italy was the leading country with 380 publications, covering 9.52% of the global total publications. In general, 49% of the global publications were contributed by developed countries such as Italy, Germany, the United Kingdom, the United States, South Korea, Spain, and Poland. Up to this point, Malaysia is the only developing country listed within the top 10 rankings.

In addition to Table 2, the distribution of countries based on the co-authorship was shown in Fig. 3. Papers co-authored by scholars from several countries are referred to as internationally collaborative articles [23], however, the benefits of this collaboration include not just expanding one's network, sharing knowledge, and improving notions, but also an effective approach for ranking higher [14]. Each point in the item density visualisation denotes colors range from white to yellow and red. For example, Germany is pointed to red because it is among the countries that have a larger number of items and higher weights (i.e., documents) compared to Northern

Table 2 Top 10 countries contributed to I4.0 publications

Country	National context	TP	%	NCP	TC	C/P	C/CP	*h*	*g*
Italy	Developed country	380	9.52	324	6360	16.74	19.63	39	64
Germany	Developed country	347	8.69	280	9437	27.20	33.70	39	92
China	Emerging and developing country	331	8.29	296	8408	25.40	28.41	42	85
United Kingdom	Developed country	284	7.11	247	7155	25.19	28.97	42	78
United States	Developed country	281	7.04	236	8239	29.32	34.91	38	86
South Korea	Developed country	262	6.56	174	1638	6.25	9.41	19	32
India	Emerging and developing country	251	6.29	196	3981	15.86	20.31	31	57
Spain	Developed country	232	5.81	197	3165	13.64	16.07	30	48
Malaysia	Developing country	207	5.18	138	1367	6.60	9.91	19	32
Poland	Developed country	170	4.26	140	1613	9.49	11.52	17	34

Notes Ranking is based on TP (Total Publication)

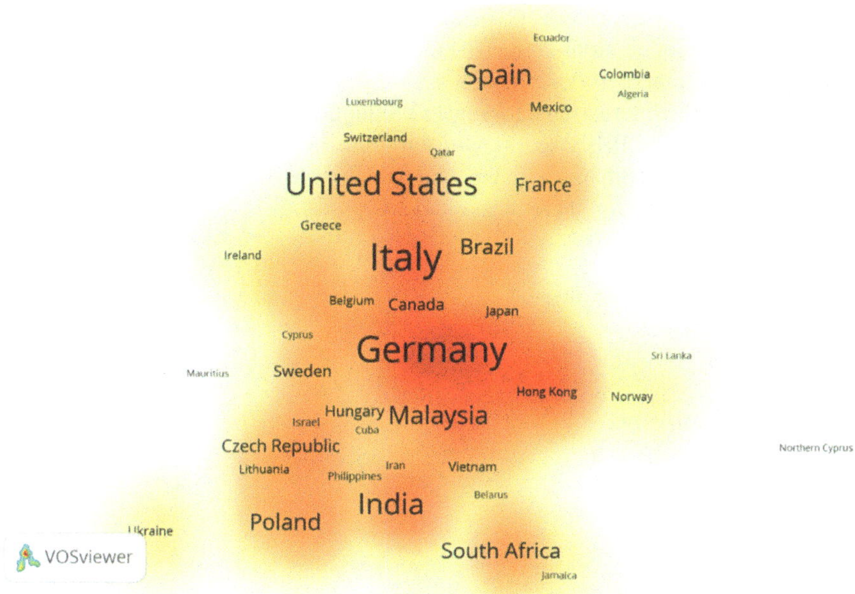

Fig. 3 Density visualisation map. *Note* An author's minimum number of documents = 1; A country's minimum number of citations = 0; 169 meets the thresholds; 68 unrelated terms deleted

Cyprus (ranked 95th). In VOSviewer (Fig. 3), the closer two countries are to one other, the more solid their bonds become [14].

According to the outcomes of co-authorship, the United Kingdom is the most connected country, linking to 63 countries through 337 co-authorships. The United States came in second (57 links, 297 co-authorships), followed by India (50, 172), Germany (49, 228), China (46, 305), and Italy (46,. 206). Yet, China had the second highest co-authorship (305 co-authorships) after the United Kingdom in terms of the total link strength. Out of 101 countries, there were 43 countries that have international collaboration (links) with less than ten countries. Though, the researchers from Albania, Trinidad and Tobago, Georgia, North Korea, and Uzbekistan were not linked to any other country for producing publications on I4.0. According to [14], research partners, international postgraduates, visiting academics, and research grants may be the criteria that contribute to international collaborations.

3.3 Core Journals, Articles, and Academic Institutions

The findings revealed that the top ten most productive journals have been issued by various publishers, as listed in Table 3. Three out of the top four journals were published by MDPI. The topmost journal was Sustainability, with 153 articles contributing 3.83% of the overall publications, followed by Applied Sciences

(111, 2.78%), and IEEE Access (107, 2.68%). However, the Applied Sciences had the lowest number of citations among the most active sources. The rest of the journals did not exceed a total of 100 publications.

Interestingly, the International Journal of Production Research (IJPR) is the second most cited journal after IEEE Access, with 3460 and 3523 total citations, respectively. Although IJPR ranked 8th based on the total publications, one of their articles written by [24] was among the top three most cited articles, with 747 citations (Table 4).

Table 3 Most proactive source titles

Source Title	Publisher	TP	%	TC	Cite score 2020	WOS quartile (Impact factor) 2020
Sustainability	MDPI AG	153	3.83	2094	3.9	Q2 (3.251)
Applied Sciences	MDPI Multidisciplinary Digital Publishing Institute	111	2.78	477	3.3	N/A
IEEE Access	Institute of Electrical and Electronics Engineers Inc	107	2.68	3523	4.8	Q2 (3.367)
Sensors	MDPI Multidisciplinary Digital Publishing Institute	70	1.75	593	5.8	Q1 (3.576)
Procedia Manufacturing	Elsevier BV	63	1.58	2789	N/A	N/A
IEEE Transactions on Industrial Informatics	IEEE Computer Society	60	1.50	1528	17.7	Q1 (10.215)
International Journal of Advanced Manufacturing Technology	Springer London	57	1.43	556	5.6	Q2 (3.226)
Technological Forecasting and Social Change	Elsevier Inc	54	1.35	1892	12.1	Q1 (8.593)
International Journal of Production Research	Taylor and Francis Ltd	52	1.30	3460	10.8	Q1 (8.568)
Computers and Industrial Engineering	Elsevier Ltd	49	1.23	1226	7.9	N/A

Table 4 Top 20 highly cited articles

No	Authors	Source	Title	Description	Cites*	Cites per Year
1	[25]	Manufacturing Letters	A Cyber-Physical Systems architecture for Industry 4.0-based manufacturing systems	Proposed a unified 5-level architecture as a guideline for implementing CPS	2085	347.50
2	[26]	Business and Information Systems Engineering	Industry 4.0	Presented the fundamental concepts of I4.0, which are described differently based on the IT-driven and the integration, automation, and decentralization of enterprise information systems	1443	206.14
3	[24]	International Journal of Production Research	Industry 4.0: State of the art and future trends	It is a survey to inform communities within industrial sectors about current advances and future prospects in the fascinating subject of I4.0	747	249.00
4	[27]	IEEE Industrial Electronics Magazine	The future of industrial communication: Automation networks in the era of the internet of things and industry 4.0	Ethernet time-sensitive networking (TSN) and the importance of 5G telecom networks in automation are the subjects of this survey	704	176.00

(continued)

Based on the CiteScore 2020 report, three journals have a CiteScore of 10 or above, which are IEEE Transactions on Industrial Informatics (17.7), Technological Forecasting and Social Change (12.1), and International Journal of Production Research (10.8). Sensors (Switzerland) is the only Q1 journal that has the lowest CiteScore that is 5.8. CiteScore, which has been an alternative to the Clarivate Analytics Impact

Table 4 (continued)

No	Authors	Source	Title	Description	Cites*	Cites per Year
5	[28]	Computer Networks	Towards smart factory for industry 4.0: A self-organized multi-agent system with big data based feedback and coordination	Provided a smart factory framework that integrates smart shop-floor items such as machines, conveyers, and goods with an industrial network, cloud, and supervisory control terminal	619	123.80
6	[29]	Computers in Industry	Industry 4.0 and the current status as well as future prospects on logistics	Identified and discussed the potential implications and pitfalls of I4.0 in the area of logistics management	554	138.50
7	[30]	IEEE Access	Peeking Inside the Black-Box: A Survey on Explainable Artificial Intelligence (XAI)	Presented a comprehensive and holistic analysis related to Explainable Artificial Intelligence (XAI)	535	178.33
8	[31]	IEEE Sensors Journal	Software-Defined Industrial Internet of Things in the Context of Industry 4.0	Analysed a software-defined IIoT architecture to identify network resource allocation and expedite information exchange mechanisms using an easily customizable networking protocol	383	76.6

(continued)

Table 4 (continued)

No	Authors	Source	Title	Description	Cites*	Cites per Year
9	[32]	SAGE Open	A Complex View of Industry 4.0	Concentrated on the relevance and impact of I4.0 and Internet-connected technologies in generating value in companies and society	372	74.40
10	[33]	IEEE Access	Digital Twin and Big Data Towards Smart Manufacturing and Industry 4.0: 360 Degree Comparison	Investigated the roles of big data and digital twin in smart manufacturing	367	122.33
11	[34]	International Journal of Production Economics	Industry 4.0 technologies: Implementation patterns in manufacturing companies	Determined distinct adoption patterns for two technology layers of I4.0—base technologies and front-end technologies. It was found that firms with a high level of implementing I4.0 tend to use most of the front-end technologies rather than a specific subset	360	180.00
12	[35]	IEEE Access	Digital Twin Shop-Floor: A New Shop-Floor Paradigm Towards Smart Manufacturing	Provided an insight into Digital Twin Shop-Floor (DTS) and a guideline for implementing four components of DTS	342	85.50

(continued)

Factor, is an indicator for journal impact measurement based on Scopus citation data. However, it should not be used as the sole metric [14]. Therefore, we included the WoS quartile (impact factor) for sources comparison. The impact factor is employed to evaluate the journals' relative importance, particularly for comparing to other journals in the same subject [23].

Table 4 (continued)

No	Authors	Source	Title	Description	Cites*	Cites per Year
13	[36]	Applied Energy	Blockchain technology in the chemical industry: Machine-to-machine electricity market	Presented two electricity producers and one electricity consumer trading with each other over a blockchain technology, which was linked to I4.0	338	84.50
14	[37]	IEEE Access	Smart Factory of Industry 4.0: Key Technologies, Application Case, and Challenges	Proposed the hierarchical architecture of a smart factory with an emphasis on essential technologies at the physical resources layer, network layer, and data application layer	328	82.00
15	[1]	International Journal of Production Economics	The expected contribution of Industry 4.0 technologies for industrial performance	Analysed the perception of the Brazilian industry about the merits of I4.0-related technologies for three industrial performance metrics—product, operational, and side-effects	299	99.67
16	[38]	International Journal of Production Research	The industrial management of SMEs in the era of Industry 4.0	Developed a framework based on the survey of 23 actual cases in which I4.0 was applied in SMEs	296	98.67

(continued)

Table 5 includes a list of the top ten most prolific universities based on the number of I4.0 papers produced. According to the World University Rankings 2022, there were no top 100 best universities mentioned. The top two institutions are Politecnico di Milano in Italy (ranked 142) and Universiti Kebangsaan Malaysia (144th). However, it seems I4.0 has not received undivided attention from the top universities.

Table 4 (continued)

No	Authors	Source	Title	Description	Cites*	Cites per Year
17	[39]	International Journal of Production Research	The impact of digital technology and Industry 4.0 on the ripple effect and supply chain risk analytics	Presented the research framework by combining the results gained from two isolated areas, i.e., the influence of digitalization on supply chain management (SCM) and the influence of SCM on the ripple effect	294	147.00
18	[40]	Journal of Manufacturing Technology Management	The future of manufacturing industry: a strategic roadmap toward Industry 4.0	Offered an integrative framework that can be served as a simple guide for the process of transition from traditional manufacturing into I4.0	284	94.67
19	[41]	ACM Computing Surveys	Cloud computing resource scheduling and a survey of its evolutionary approaches	Presented the taxonomy of managing and scheduling cloud resources	280	46.67
20	[42]	International Journal of Production Research	A dynamic model and an algorithm for short-term supply chain scheduling in the smart factory industry 4.0	Proposed a model and algorithm for short-term supply chain scheduling in smart factories I4.0	276	55.20

3.4 Author's Contribution to the Development of the Topic

Table 6 lists the 10 most prolific authors, who cover 3.14% of total publications in the area of I4.0. The authors were affiliated with seven countries, as follows: four authors from China and one author each from Austria, Italy, Spain, Brazil, Thailand, and Germany. The top two authors—Li, D. and Wan, J.—were from the South China University of Technology. Even though Wan, J. has two fewer publications than Li, D.; yet, he holds the most total citations of 2288.

Table 5 Most prolific universities with a minimum of five publications

Affiliation	Rankings[a]	Country	TP	%	NCP	TC	C/P	C/CP	*h*	*g*
University of Johannesburg	=434	South Africa	47	1.18	32	331	7.04	10.34	8	17
Universiti Kebangsaan Malaysia	144	Malaysia	37	0.93	23	80	2.16	3.48	5	7
Politecnico di Milano	142	Italy	37	0.93	32	1213	32.78	37.91	15	32
Universidad Politécnica de Madrid	=459	Spain	30	0.75	27	329	10.97	12.19	10	17
Silesian University of Technology	801–1000	Poland	28	0.70	20	171	6.11	8.55	8	12
University of Naples–Federico II	=424	Italy	28	0.70	25	358	12.79	14.32	10	18
South China University of Technology	=407	China	28	0.70	27	2448	87.43	90.67	17	27
Universiti Teknologi Malaysia	=191	Malaysia	27	0.68	19	146	5.41	7.68	7	11
Free University of Bozen-Bolzano	651–700	Italy	26	0.65	23	412	15.85	17.91	12	19
University of the Basque Country	701–750	Spain	26	0.65	24	459	17.65	19.13	8	21

[a] 2022 World University Rankings: https://www.topuniversities.com/university-rankings/world-university-rankings/2022

Table 6 Most prolific authors

Author's Name	Author's Affiliation & Country	TP	%	NCP	TC	C/P	C/CP	*h*	*g*
Li, D	School of Mechanical and Automotive Engineering, South China University of Technology, Guangzhou, China	18	0.45	18	2022	112.33	112.33	14	18
Wan, J	School of Mechanical and Automotive Engineering, South China University of Technology, Guangzhou, China	16	0.40	16	2288	143.00	143.00	14	16
Müller, J.M	Logistics and Operations Management, Salzburg University of Applied Sciences, Puch bei Hallein, Austria	14	0.35	13	1057	75.50	81.31	9	13
Rauch, E	Faculty of Science and Technology, Free University of Bozen-Bolzano, Bolzano, Italy	14	0.35	13	197	14.07	15.15	8	13
Fraga-Lamas, P	Department of Computer Engineering, Universidade da Coruña, Coruña, Spain	11	0.28	11	445	40.45	40.45	8	11
Tortorella, G.L	Universidade Federal de Santa Catarina, Florianópolis, Brazil	11	0.28	10	454	41.27	45.40	7	10
Wang, S	School of Mechanical and Automotive Engineering, South China University of Technology, Guangzhou, China	11	0.28	11	1687	153.36	153.36	9	11
Jermsittiparsert, K	Social Research Institute, Chulalongkorn University, Bangkok, Thailand	10	0.25	9	241	24.10	26.78	3	9
Tao, F	School of Automation Science and Electrical Engineering, Beihang University, Beijing, China	10	0.25	9	1266	126.60	140.67	8	9
Voigt, K.I	Friedrich-Alexander-University Erlangen-Nürnberg, Nürnberg, Germany	10	0.25	10	962	96.20	96.20	7	10

The most cited article (619 citations), which has been co-authored by both Li and Wan, i.e., Wang et al. [28], was published in the Computer Networks journal. This study presents a model to build up smart shop-floor objects such as machines, conveyers, and products as agents, as well as an intelligent negotiation mechanism for them to collaborate. They believe this industrial network will enable system-wide feedback and coordination using big data analytics to optimise system performance.

3.5 Common Terminology and Distribution of I4.0 Publications Based on Major Technologies

A total of 60 author keywords with a minimum of 15 occurrences, were recorded for the mapping in VOSviewer. The findings were derived after renaming congeneric phrases and synonymic single words from a total of 10,058 keywords. For example, "4th Industrial Revolution" (17 occurrences, 9 links), "fourth industrial revolution" (146 occurrences, 46 links), "Industrial Revolution" (41 occurrences, 28 links), "Industrial Revolution 4.0" (30 occurrences, 16 links), "Industry 4.0" (20 occurrences, 28 links), "Industry 40" (15 occurrences, 16 links), and "the fourth industrial revolution" (25 occurrences, 20 links) were generalized with the "Industry 4.0" (1652 occurrences, 103 links). Our results showed that "Industry 4.0" is the most often observed keyword, with 1886 occurrences and 59 links (Fig. 4). IoT and CPS, which are the two fundamental components of I4.0, were ranked second and fourth, with 390 and 239 occurrences, respectively.

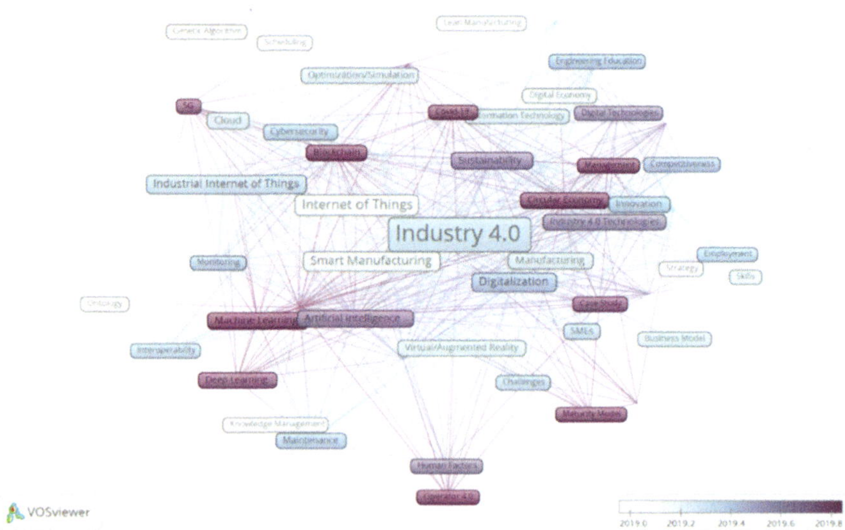

Fig. 4 Overlay visualisation map of a term co-occurrence network

The outbreak of Coronavirus-19 (Covid-19) has made this keyword the latest term with an average publication of 2020.393 (Fig. 4). Importantly, Covid-19 is revealing deep interconnections between pandemics and "industry 4.0" (17 link strength), "IoT" (4 link strength), and "industry 4.0 technologies" (3 link strength) in unforeseen ways. Moreover, "circular economy", which is an emerging concept mainly in the area of eliminating wastes, was among the top three keywords with the latest average publication year of 2020.

In addition, we noticed several distributions of I4.0 publications with other nine major technologies, which are Big Data Analytics, Industrial IoT, Simulation/Optimization, Additive Manufacturing, Horizontal/Vertical System Integration, Virtual/Augmented Reality, Autonomous Robots, Cloud, and Cybersecurity—see Table 7. I4.0 has the highest link strength with all nine major technologies. The most link strength recorded is between I4.0 and IoT (241 links strength). However, there is no link strength documented between AM and optimization/simulation, AM and CPS, cybersecurity and VR/AR, and cybersecurity and AM. It is also interesting to see the connection between I4.0 and other keywords such as sustainability (72 links strength), circular economy (38 links strength), lean manufacturing/production (32 links strength), sustainable development (16 links strength), and sustainable manufacturing/production (8 links strength). Surprisingly, there is no network of co-occurrences for green manufacturing/production. Delving into these technologies, the impact of some of them on developing the manufacturing's integral elements—products, processes, and systems—have been potentially exposed [3, 4]:

- *Big Data Analytics*–capable of improving direct/indirect costs, waste and emissions, and product end-of-life management at the product level [43–45], energy consumption and environmental impact at the process level [46–48], net profit,

Table 7 Link strength of nine major technologies in I4.0 publications

Label	Keywords	Link strength										
		A	B	C	D	E	F	G	H	I	J	K
A	Big Data Analytics (BDA)											
B	Optimization/Simulation	1										
C	Cloud	19	3									
D	Virtual/Augmented Reality (VR/AR)	6	1	2								
E	Cyber-Physical Systems (CPS)	30	4	23	6							
F	Industrial Internet of Things (IIoT)	15	1	24	11	24						
G	Additive Manufacturing (AM)	4	–	1	4	–	1					
H	Autonomous Robots	6	2	1	4	5	5	5				
I	Cybersecurity	4	1	5	–	9	10	–	2			
J	Internet of Things (IoT)	53	5	36	11	75	26	11	14	12		
K	Industry 4.0	97	32	71	38	158	93	29	50	17	241	

operational performance, material use & efficiency, energy use & efficiency, and water use & efficiency at the system level [7, 48];

- *Virtual and Augmented Reality*–capable of improving product quality & durability, functional performance, and safety & health impact at the product level [47, 49], manufacturing cost, personnel health, and operational safety at the process level [7, 50], net profit, operational performance, health & safety, and stakeholder engagement at the system level [51],
- *Optimization and Simulation*–capable of improving functional performance at the product level [52], manufacturing cost, energy consumption, environmental impact, personnel health, and operational safety at the process level [45, 47], capital charge, manufacturing cost, operational performance, material use & efficiency, energy use & efficiency, water use & efficiency, waste and emission, and stakeholder engagement at the system level [7, 53, 54],
- *Additive Manufacturing*–capable of improving initial investments, material use & efficiency, energy use & efficiency at the product level [2, 7, 43, 55], personnel health and operational safety at the process level [7], net profit, operational performance, health & safety, and stakeholder engagement at the system level [45],
- *Cloud*–capable of improving functional performance, product end-of-life management, and safety & health impact at the product level [34], manufacturing cost and waste management at the process level [47], net profit, manufacturing cost, operational performance, health & safety, and stakeholder engagement at the system level [45, 56, 57],
- *Industrial Internet of Things*–capable of improving benefits and losses, product quality and durability, and product end-of-life management at the product level [34, 49, 58], manufacturing cost, waste management, personnel health, operational safety at the process level [7, 44, 46, 52], net profit, capital charge, operational performance, health & safety, and stakeholder engagement at the system level [44, 46, 56].

4 Limitations and Conclusion

Notwithstanding its contribution, this investigation is subject to the limitations that may generate future research directions. First, although this study covers I4.0 research up to May 2021, the subject is anticipated to expand further, necessitating subsequent bibliometric and network analyses of the I4.0 literature. Despite the recognised theoretical overlap across some of the observed I4.0 themes and technologies, it is critical to rerun the presented analyses within the next few years once I4.0 studies have further evolved, so as to compare the new findings to those provided in this work. In this regard, complementary or alternate approaches such as correspondence analysis, data clustering, etc. may also be applied. The performed bibliometric structure, which could alter when new I4.0 publications are released, may also be tested by researchers. Next, the documented analyses are limited to the Scopus database, thus overlooking other databases such as the Web of Science may give differing insights. Furthermore, the current study was restricted to journal articles, thus scholars could also conduct bibliometric and network analyses of I4.0 studies based on more than just journal articles.

In conclusion, this article has mapped out the methodological structure of the I4.0 literature using Bibliometric analysis in order to demonstrate how the process of the diffusion and adoption of I4.0 concepts and technologies, which typically flows from developed countries, has scientifically progressed. Italy with 380 publications was ranked first among the contributing countries, covering 9.52% of the overall publications. In general, 49% of the global publications were contributed by developed countries including Italy, Germany, the United Kingdom, the United States, South Korea, Spain, and Poland. Malaysia was the only developing country listed within the top 10 rankings. Based on the identified patterns, prolific authors and institutions as well as the major technologies related to Industry 4.0 were also discussed. Even though this study describes a ground-breaking bibliometric and network analysis concerning the I4.0 scholarship, the field's dynamic nature necessitates ongoing screening and mapping in the future years.

References

1. Dalenogare LS, Benitez GB, Ayala NF, Frank AG (2018) The expected contribution of Industry 4.0 technologies for industrial performance. Int J Prod Econ 204:383–394
2. Stock T, Seliger G (2016) Opportunities of sustainable manufacturing in industry 4.0. Proc CIRP 40:536–541
3. Enyoghasi C, Badurdeen F (2021) Industry 4.0 for sustainable manufacturing: Opportunities at the product, process, and system levels. Resour Conserv Recycl 166:105362
4. Gholami H, Abu F, Lee JKY, Karganroudi SS, Sharif S (2021) Sustainable manufacturing 4.0—pathways and practices. Sustainability 13(24):13956
5. Rüßmann M, Lorenz M, Gerbert P, Waldner M, Justus J, Engel P, Harnisch M (2015) Industry 4.0: the future of productivity and growth in manufacturing industries. Boston Consult Group 9(1):54–89
6. Sartal A, Bellas R, Mejías AM, García-Collado A (2020) The sustainable manufacturing concept, evolution and opportunities within Industry 4.0: a literature review. Adv Mech Eng 12(5):1–17
7. Wee D, Kelly R, Cattel J, Breunig M (2015) Industry 4.0-how to navigate digitization of the manufacturing sector. McKinsey Company:58
8. Kagermann H, Wahlster W, Helbig J (2013) Recommendations for Implementing the Strategic Initiative Industrie 4.0. Final report of the Industrie 4.0 WG
9. Bernat S, Karabag SF (2019) Strategic alignment of technology: Organising for technology upgrading in emerging economy firms. Technol Forecast Soc Chang 145:295–306
10. Comin D, Hobijn B (2004) Cross-country technology adoption: making the theories face the facts. J Monet Econ 51:39–83
11. Lee JKY, Gholami H, Saman MZM, Ngadiman NHAB, Zakuan N, Mahmood S, Omain SZ (2021) Sustainability-oriented application of value stream mapping: a review and classification. IEEE Access 9:68414–68434
12. Pauna VH, Picone F, Guyader GL, Buonocore E, Franzese PP (2018) The scientific research on ecosystem services: a bibliometric analysis. Ecol Quest 29(3):53–62

13. Abu F, Gholami H, Saman MZM, Zakuan N, Sharif S, Streimikiene D (2021) Pathways of lean manufacturing in wood and furniture industries: a bibliometric and systematic review. Eur J Wood Wood Prod 79(4):753–772
14. Khudzari JM, Kurian J, Tartakovsky B, Raghavan GSV (2018) Bibliometric analysis of global research trends on microbial fuel cells using Scopus database. Biochem Eng J 136:51–60
15. Letchumanan LT, Gholami H, Yusof NM, Ngadiman NHAB, Salameh AA, Štreimikienė D, Cavallaro F (2022) Analyzing the factors enabling green lean six sigma implementation in the Industry 4.0 Era. Sustainability 14(6):3450
16. Piwowar-Sulej K, Krzywonos M, Kwil I (2021) Environmental entrepreneurship–Bibliometric and content analysis of the subject literature based on H-Core. J Clean Prod:295
17. Adam A, Zakuan NB, Gholami H, Bichi AA, Uba SA (2019) Trends in Industry 4.0 research: a bibliometric analysis of publications in internet of thing component. Int J Recent Technol Eng 8(3S2):766–775
18. Egghe L (2006) Theory and practise of the g-index. Scientometrics 69(1):131–152
19. Harzing AW (2007) Publish or Perish. Retrieved from https://harzing.com/resources/publish-or-perish
20. Hirsch JE (2005) An index to quantify an individual's scientific research output. In: Paper presented at the Proceedings of the national academy of sciences of the United States of America
21. Van Eck NJ, Waltman L (2020) Manual for VOSviewer version 1.6.16
22. Ziesing D, Hochstein N (2013) Engineering tools as a basis for industry 4.0. ThyssenKrupp techforum 1:72–77
23. Zeng J, Qu J, Ma H, Gou X (2021) Characteristics and trends of household carbon emissions research from 1993 to 2019: a bibliometric analysis and its implications. J Clean Prod:295.
24. Xu LD, Xu EL, Li L (2018) Industry 4.0: State of the art and future trends. Int J Prod Res 56(8):2941–2962
25. Lee J, Bagheri B, Kao HA (2015) A cyber-physical systems architecture for Industry 4.0-based manufacturing systems. Manuf Lett 3:18–23
26. Lasi H, Fettke P, Kemper HG, Feld T, Hoffmann M (2014) Industry 4.0. Bus Inf Syst Eng 6(4):239–242
27. Wollschlaeger M, Sauter T, Jasperneite J (2017) The future of industrial communication: Automation networks in the era of the internet of things and industry 4.0. IEEE Ind Electron Mag 11(1):17–27
28. Wang S, Wan J, Li D, Zhang C (2016a) Implementing smart factory of industrie 4.0: an outlook. Int J Distrib Sens Netw 12(1):3159805
29. Hofmann E, Rüsch M (2017) Industry 4.0 and the current status as well as future prospects on logistics. Comput Ind 89:23–34
30. Adadi A, Berrada M (2018) Peeking inside the black-box: a survey on explainable artificial intelligence (XAI). IEEE Access 6:52138–52160
31. Wan, J., Tang, S., Shu, Z., Li, D., Wang, S., Imran, M., & Vasilakos, A.V. (2016). Software-Defined Industrial Internet of Things in the Context of Industry 4.0. *IEEE Sensors Journal*, 16(20), 7373–7380.
32. Roblek V, Meško M, Krapež A (2016) A complex view of industry 4.0. SAGE Open 6(2)
33. Qi Q, Tao F (2018) Digital twin and big data towards smart manufacturing and Industry 4.0: 360 degree comparison. IEEE Access 6:3585–3593
34. Frank AG, Dalenogare LS, Ayala NF (2019) Industry 4.0 technologies: Implementation patterns in manufacturing companies. Int J Prod Econ 210:15–26
35. Tao F, Zhang M (2017) Digital twin shop-floor: a new shop-floor paradigm towards smart manufacturing. IEEE Access 5:20418–20427
36. Sikorski JJ, Haughton J, Kraft M (2017) Blockchain technology in the chemical industry: machine-to-machine electricity market. Appl Energy 195:234–246
37. Chen B, Wan J, Shu L, Li P, Mukherjee M, Yin B (2017) Smart factory of Industry 4.0: key technologies, application case, and challenges. IEEE Access 6:6505–6519

38. Moeuf A, Pellerin R, Lamouri S, Tamayo-Giraldo S, Barbaray R (2018) The industrial management of SMEs in the era of Industry 4.0. Int J Prod Res 56(3):1118–1136
39. Ivanov D, Dolgui A, Sokolov B (2019) The impact of digital technology and Industry 4.0 on the ripple effect and supply chain risk analytics. Int J Prod Res 57(3):829–846
40. Ghobakhloo M (2018) The future of manufacturing industry: a strategic roadmap toward Industry 4.0. J Manuf Technol Manag 29(6):910–936
41. Zhan ZH., Liu XF, Gong YJ, Zhang J, Chung HSH, Li Y (2015) Cloud computing resource scheduling and a survey of its evolutionary approaches. ACM Comput Surv 47(4)
42. Ivanov D, Dolgui A, Sokolov B, Werner F, Ivanova M (2016) A dynamic model and an algorithm for short-term supply chain scheduling in the smart factory industry 4.0. Int J Prod Res 54(2):386–402
43. Nascimento DLM, Alencastro V, Quelhas OLG, Caiado RGG, Garza-Reyes JA, Rocha-Lona L, Tortorella G (2019) Exploring Industry 4.0 technologies to enable circular economy practices in a manufacturing context: a business model proposal. J Manufact Technol Manag 30(3):607–627
44. Rajput S, Singh SP (2019) Industry 4.0−challenges to implement circular economy. Benchmark Int J 28(5):1717–1739
45. Song Z, Moon Y (2017) Assessing sustainability benefits of cybermanufacturing systems. Int J Adv Manufact Technol 90(5–8):1365–1382
46. Blunck E, Werthmann H (2017) Industry 4.0–an opportunity to realize sustainable manufacturing and its potential for a circular economy. DIEM 3(1):644–666. https://hrcak.srce.hr/187419
47. Carvalho ACP, Carvalho APP, Carvalho NGP (2020) Industry 4.0 technologies: what is your potential for environmental management?. In: Industry 4.0 current status and future trends. IntechOpen
48. Dubey R, Gunasekaran A, Childe SJ, Papadopoulos T, Luo Z, Wamba SF, Roubaud D (2019) Can big data and predictive analytics improve social and environmental sustainability? Technol Forecast Soc Chang 144:534–545
49. Tao F, Cheng J, Qi Q, Zhang M, Zhang H, Sui F (2018) Digital twin-driven product design, manufacturing and service with big data. Int J Adv Manuf Technol 94(9–12):3563–3576
50. Menon S, Shah S, Coutroubis A (2018) Impacts of I4. 0 on sustainable manufacturing to achieve competitive advantage. In: Paper presented at the 8th International Conference on Operations and Supply Chain Management. Cranfield, UK
51. Rumsey A, Le Dantec CA (2020) Manufacturing change: the impact of virtual environments on real organizations. In: Paper presented at the Proceedings of the 2020 CHI Conference on Human Factors in Computing Systems
52. MacArthur E, Waughray D (2016) Intelligent assets: unlocking the circular economy potential. Ellen MacArthur Foundation, Cowes, UK
53. Hashemi A, Gholami H, Venkatadri U, Salameh AA, Jafari M, Abdul-Nour G (2022) A novel approach to solve cell formation problems with alternative routing using particle swarm optimization, Transf Bus Econ 21(55):313–332
54. Hashemi A, Gholami H, Venkatadri U, Sattarpanah Karganroudi S, Khouri S, Wojciechowski A, Streimikiene D (2021) A new direct coefficient-based heuristic algorithm for set covering problems. Int J Fuzzy Syst 24:1131–1147
55. Kamble SS, Gunasekaran A, Gawankar SA (2018) Sustainable Industry 4.0 framework: a systematic literature review identifying the current trends and future perspectives. Process Saf Environ Prot 117:408–425
56. Esmaeilian B, Sarkis J, Lewis K, Behdad S (2020) Blockchain for the future of sustainable supply chain management in Industry 4.0. Resour Conserv Recycl 163:105064
57. Ivanov D (2018) New drivers for supply chain structural dynamics and resilience: sustainability, Industry 4.0, self-adaptation. In: Structural dynamics and resilience in supply chain risk management. Springer, pp 293–313
58. Wang S, Wan J, Zhang D, Li D, Zhang C (2016b) Towards smart factory for industry 4.0: a self-organized multi-agent system with big data based feedback and coordination. Comput Netw 101:158–168

Smart Laser Welding: A Strategic Roadmap Toward Sustainable Manufacturing in Industry 4.0

Ahmad Aminzadeh, Davood Rahmatabadi, Mostafa Pahlavani, Mahmoud Moradi, and Jonathan Lawrence

1 Introduction

Industry 4.0 is the fourth significant period in the industry since the beginning of the Industrial Revolutions. It defines a way to make the transition from dominant machine production to digital production. It transforms the production and management systems in every industry and every country into smart ones. In short, connecting the digital world to the physical aspect of the industry using the growing capabilities of the Internet of Things (IoT) and other technologies can be described as the Fourth Industrial Revolution [1]. On the other hand, this course associates with the emergence of new technologies in robotics, artificial intelligence, blockchain, nanotechnology, quantum processing, biotechnology, the Internet of Things, and automobiles, which can facilitate their launch, further prosperity, and expansion in various fields. The concept of "Industry 4.0" has recently been introduced and accepted in

A. Aminzadeh
Department of Mathematics, Computer Science and Engineering, University of Quebec at Rimouski, Rimouski, Quebec, Canada

D. Rahmatabadi
School of Mechanical Engineering, University of Tehran, Tehran, Iran

M. Pahlavani
Department of Research and Development, MAPNA Rail Technologies (MRT) Company, Karaj, Iran

School of Automotive Engineering, Iran University of Science and Technology, Tehran, Iran

M. Moradi (✉)
Faculty of Arts, Science and Technology,
University of Northampton, Northampton NN1 5PH, UK
e-mail: Mahmoud.Moradi@northampton.ac.uk

J. Lawrence
School of Computing, Engineering & Physical Sciences, University of the West of Scotland (UWS), High Street, Paisley PA1 2BE, UK

H. Gholami et al. (eds.), *Sustainable Manufacturing in Industry 4.0*,
https://doi.org/10.1007/978-981-19-7218-8_3

the academic world and the manufacturing community. However, the shift from the third industrial revolution to the world of Industry 4.0 demands comprehensive study toward understanding irreversible transformations. The standards of this new industrial revolution need to be well understood, and a clear roadmap must be developed and implemented to achieve a successful change. There are various elements and factors in these changes that also stimulate social influences. The Internet of Things is one of these instrumental requirements that allows machines to communicate and creates a cheaper production environment.

The second important motivation and advantage of these transformations is "automation" in those systems act only to be influenced by each other's performance [2]. On the other hand, some sensors [3] and cyber-physical systems (CPS) are other critical parts of this evolution and provide accessible communication between machines and surrounding environment. When different aspects of Industry 4.0, including CPS, IoT, machine-to-machine (M2M) communications, and automation, come together, it becomes easier to build compatible, robust, and agile systems with exceptional capacities. This transformation enables the communication of machines with human operators, leading to creating a new construction vision based on four fundamental concepts of 1. intelligence, 2. products, 3. communication, and 4. information network.

Like the other three industrial revolutions, the fourth industrial revolution has affected various manufacturing industries, including welding processes, and this has become important for achieving the desired efficiency and effectiveness in production. It is natural that manufacturers that cannot immediately adapt to modern market demand either go out of business or incur higher costs in the future.

Nowadays, welding is an advanced technique that is used in all aspects of industries and modern life. Laser welding (LW) has been an essential technology in numerous businesses such as IT, manufacturing, health care, and beauty since Einstein's establishment of the theoretical foundations in 1917. With the development of laser applications in material processing methods, laser welding has always been considered a new process in various industries [4–6]. Technologies such as scanning and deposition welding have led to the development of laser welding methods in multiple sectors such as the automotive and aerospace industries. Although in the '60s there were attempts to use lasers in welding, the '70s can be considered the beginning of laser welding in various industries. Solid-state pulsed lasers were the first lasers used in laser welding for applications such as spot-on sensitive electronic or precision mechanical components. With the progress of laser beam production resources and technological advances in laser beam conduction, laser welding has also undergone dramatic changes. With the increasing use of sheet metal industries, new technologies to improve product quality and reduce costs and operating time have become significant. Since welding plays a vital role in producing metal products, new laser welding methods instead of traditional welding methods increase the quality of products. It also reduces the number of operations after the welding process in various industries. Recently, Aminzadeh et al. [7] used a real-time monitoring technique to define the distortion and deviation in aluminum laser-welded blanks via a 3D laser scanner.

For this reason, this process is prevalent in the industry and is an attractive option. Although laser welding imposes complex processes in process control and data connectivity, it offers numerous unparalleled advantages like speed, technology, and costs. Compared with traditional fabrication processes, including resistance spot welding or conventional ARC welding, laser welding or primarily fiber laser welding provides easy operation. It is straightforward to learn with a fast-learning curve, improves energy efficiency and machine lifetime. It causes a smaller ecological footprint, more economical maintenance requirements, less environmental contamination, and less high-volume production time. However, a lack of a road map in the Twins model, metaheuristic approaches, and machine learning toward potential advancements is a gray area in manufacturing sectors. Therefore, this chapter aims at introducing the use of big data (BD) and AI-ML in designing digital twins (DTs) or DT-based systems for laser welding applications by highlighting the current state-of-the-art deployments. Finally, a comprehensive comparison of sustainability factors for different welding processes has been reviewed.

2 Real-Time Monitoring for Smart Welding

The technology of Laser Welding (LW) as a permanent connection technique has notable potential for industrial applications. As compared to conventional welding methods, LW shows advantages of productivity, versatility, effectiveness, deeper penetration, less distortion, and higher welding velocity [8]. However, LW is an almost complex fabrication process in which achieving acceptable joint quality is affected by several process variables and other factors such as defects in the microstructure of the base material, contamination of the workpiece surface, and changes in laser beam properties. Such defects alter the welded elements' mechanical characteristics, resulting in an increased risk of fatigue of the part. Therefore, ensuring the quality of the welded joint is a must for use in industry. For this purpose, quality monitoring is considered critical in modern production systems and is usually applied in three stages before, during, and after the process [9].

Since laser welding inputs (controllable and uncontrollable), noises, and disturbances alongside the welding circumstances affect the outputs, a promising method of producing the desired results is in demand to guarantee that all noises are dismissed or lessened; therefore, the welding process conditions maintain at their standard levels.

One way of ensuring the mentioned conditions is monitoring the principal inputs means welding parameters and other influential welding conditions that may deviate from standard levels/constants or even change completely. Simple changes would be enough for some deviations of welding conditions or input parameters from desired levels or constants. For instance, adjusting the torch position can remove the error due to deviation between the actual weld seam and the torch's trajectory. Or, if welding's current becomes lower than its desired value, action will be taken to increase its level to the needed constants. These mentioned corrections are straightforward because they are applied just to some individual welding conditions and variables.

However, some errors are related to the complex difficulties from unknown sources and cannot be corrected by changing individual parameters; therefore, dealing with these conditions needs smart and in-process control and monitoring.

2.1 Intelligent Laser Welding Monitoring

Due to its smaller heat-affected zone, higher operation speed, and precision, LW has recently become a widely employed joining technology in different industries, ranging from the assembly line of the vehicle's production to micro-electric elements in the electronic industries. Faster, reliable, and economical detection of defects in industry and manufacturing processes is one of the chief concerns. Therefore, several developed systems with the ability of online inspection have recently been introduced to improve welded components' quality plus reduce costs. Generally, non-destructive weld quality monitoring is classified into three main categories: 1. pre-process, 2. post-process, and 3. in-process [10, 11]—see Fig. 1 [12].

Pre-process inspection is ideal for adjusting the system prior to starting the welding; however, there are some limitations due to part fit-up problems. This scanning process focuses mainly on the issues of seam tracking and the gaps between the components that will be welded to each other. The pre-process monitoring ensures a reliable joint by adjusting the beam spot to focus on the center gaps. In other words, preconditions assure a correct weld line position and, therefore, a sound welding condition.

In welding with smaller, higher speed, and advanced laser beam spots that are more accurate, the precise position of the beam is more vital. For this purpose, seam tracking sensors with the ability to operate both offline and in a closed control loop can be used to inspect and detect the seam position and helps operators in modifying the part dimensions, weld line position, and clamping tolerances.

Second, post-process inspection focuses on the final product's global quality before its delivery to the customer. Although some problems of the produced elements

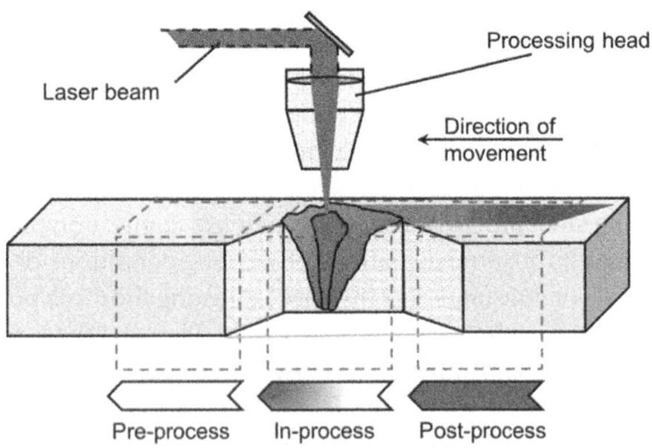

Fig. 1 Three monitoring stages

can be corrected by using post-processing inception, this detection method is sometimes considered distractive because it cannot recover the defective components that have already been fabricated.

In this regard, modern manufacturing systems are considered online monitoring as in-process inspection during laser welding. Adding a quality control system to a laser welding machine increases productivity by quickly identifying and reducing the resulting number of defects in a process. Process monitoring can support the identification of defective welds. However, as it often lacks complete reliability, such detection is usually only indicative. Some of the defects are easy to detect online during the process, mainly if generated at the surface, while others are challenging to observe, particularly internal imperfections. To define the process condition correctly, characterizing and discovering different types of welding defects is necessary. In this chapter, online monitoring is considered a core of sustainability in laser material processing. Three classifications of the monitoring process, which were mentioned in this section, are shown in Table 1.

Table 1 Three monitoring stages, equipment, signals, and objectives [13]

Stages	Equipment	Monitoring signals	Objective
Pre-process	Machine vision Laser triangulation	Optical signal	Seam tracking Gap measuring
In-process	Fusion sensors Laser 3D scanners	Optical signal Acoustic signal Electrical signal Thermal signal Ultrasonic signal	Welding stability Defect detection Pool monitoring Keyhole monitoring Feature prediction Feedback control
Post-process	Machine vision NDT methods Metallurgical test Laser triangular	Optical signal Acoustic emission	Defect classification Weld geometry

2.2 *In-Process Monitoring for Sustainable Manufacturing*

Online/in-process monitoring is the first step for achieving cloud manufacturing and an intelligent decision base for each system. The ever-increasing demands for a product with improved characteristics and quality produced at higher rates put an undue burden on online monitoring improvements during the automatic welding process. With no online monitoring devices and sensors, flaws will remain undetected, which can cause costly correction or repair situations.

Recently, some employed devices, such as laser-triangulation cameras, help operators with online and in situ control and monitoring the weld bead dimension and geometry [14].

This inspecting/monitoring method has been successfully applied in several sheet metal processes, including vehicles' production lines or tailor-welded blanks [6, 7]. However, this procedure is not capable of detecting internal flaws, and therefore cannot be employed in some of the fabrication processes, like remote laser welding. Other monitoring techniques, such as coaxial, optical radiation detection [15], coaxial visual detection [16], paraxial sound [17], temperature detection [18], plasma radiation, and charge detection [19], have been developed for Laser Process Control System (LPCS) that can be used alone or in combination with 3D laser triangulation inspection camera.

As mentioned before, the data transition is the linking chain in any intelligent system, and in this regard, the Internet of Things sensor plays a vital role in smart factories. With the ability to transfer real-time data and take corrective and preventive measures instantly, IoT helps to reduce the maintenance time, facilitate online inspection and monitoring, and optimization of production systems.

3 Robots in Welding

Technavio Research has recently reported the day-by-day growth of robotics due to the development of Industry 4.0. The concept of robotics in Industry 4.0 is creating intelligent industrial sectors by which production lines would benefit from smart systems and IoT and avoid disturbances [20]. Robots are beneficial in operating different tasks from monitoring the machines' conditions, analyzing, diagnosing, and predicting the failures to moving heavy objects [21]. They are also helpful in assembling products, treating dangerous materials, painting and cutting and shaping warehouses, polishing, etc. Kuka robots are examples of industrial robots used in material treatment, loading or unloading, spot, laser and arc welding, and palletizing or depalletizing. They are capable of delivering real-time data through IoT. The fast development of smart manufacturing with the help of artificial intelligence, known as AI, makes the traditional offline programming and teaching-playback modes obsolete since they cannot adapt themselves to the flexible and fast modes of modern manufacturing. In this regard, intelligent and automated industrial welding robots have been introduced and added to production lines with the aim of enhancing efficiency.

4 Smart Decisions in Manufacturing

Too often, intelligence factories leverage the universal shop floor data collection provided by intelligent manufacturing equipment, connected sensors, and a wealth of IoT devices to improve the performance of industrial operations significantly. Nevertheless, data transmission and big computing involve reorganizing the entire organization as operations and information technology congregate.

Traditionally, C-suite executives think about better and faster decision-making because from their point of view, it's up to them, the leaders. However, now, smart manufacturing's ability to deliver data efficiently and quickly to the decision point is changing such long, slow-moving traditional trends of decision-making. Sustainability is defined as ongoing growth in economic and social fields with a less negative impact on the environment.

Intelligent manufacturing ensures obtaining all of the three aspects of sustainable development together with the help of Industry 4.0 components, including real-time transferring data, reducing waste and the loss of material, enhancing productivity, cutting down costs and the time of processes, and dealing with the labor force challenges. In modern manufacturing systems, a researcher uses metaheuristic approaches and artificial intelligence in divers' applications to define and optimize the influence of process parameters.

5 Digital Twins, Big Data, and Connection Interfaces in Laser Welding 4.0

Recently, the digital twin or device shadow as a part of Industry 4.0 is getting more and more attention between all industries and academic fairs. More precisely, digital twinning is one of the top ten technology trends in the last couple of years due to its high applicability in the industrial sector. As this trend unfolds, manufacturing processes play a crucial role and are becoming increasingly digital. Generally, a digital twin, as a link between digital models and simulations with real-world data, creates new possibilities for improved creativity, competitive advantage, and human-centered design. A complete real-time presentation of the state of the intelligent manufacturing system is a challenge; however, the emergence of a digital twin has made it possible to solve this problem. Undeniably, digital twin solutions, a near-real-time digital image of a physical object, and real-time monitoring are the most common in the ear of intelligent technologies. Frequently, it is considered as part of the intelligent fabrication

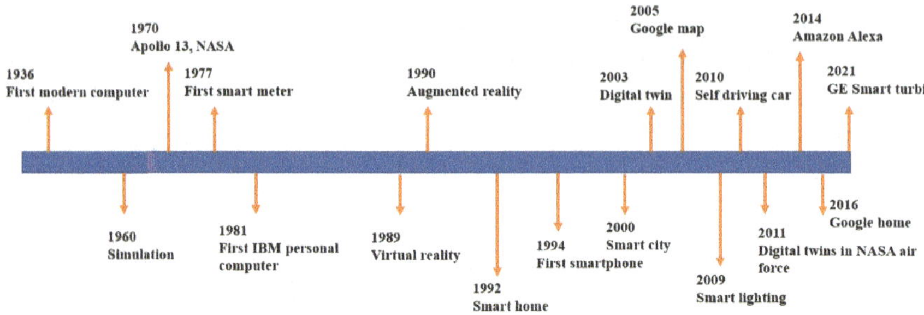

Fig. 3. Evolutionary trend of digital twin concept

process. However, it can be employed in any field, such as production, training and education, marketing and business, transportation, energy, power, electronics, human and healthcare, sports and games, networking, and communications. Production systems can observe, monitor, and control physical processes, produce a digital twin within the physical world, obtain and collect real-time data/information from the surrounding environments, analyze and simulate the conditions, and finally make decisions based on real-time communication and collaboration with humans. Integrating the digital twin into intelligent manufacturing makes construction processes smarter, efficient, and more available. Sustainable and smart manufacturing includes sustainable, intelligent manufacturing facilities, systems, and services assisting and supporting each other. Smart manufacturing equipment has two dimensions: intelligent manufacturing unit and line. A twin model unit and the line could be simulated in real-time condition, and all reports are considered in the design, production, logistics, and sales. The following diagram, Fig. 3, shows the advanced progress in the digital twins' concept, especially in manufacturing science and technology.

6 Intelligent Manufacturing

In the Industry 4.0 age, smart production and manufacturing processes have drawn attention due to the demand for sustainable development. Smart manufacturing has to consider sustainability features. Its facilities, such as laser beam welding and industrial robots, should be more intelligent, supporting their combination toward the smart manufacturing closed loop to perform different tasks. The systems of intelligent manufacturing show a diversified trend, and a growing number of them are being developed for particular responsibilities and applied to actual products; therefore, they can significantly improve the level of intelligence. Current services in intelligent manufacturing are investigated and developed, and the sustainable collaborative manufacturing system platform integrates consumers, specialists, and businesses

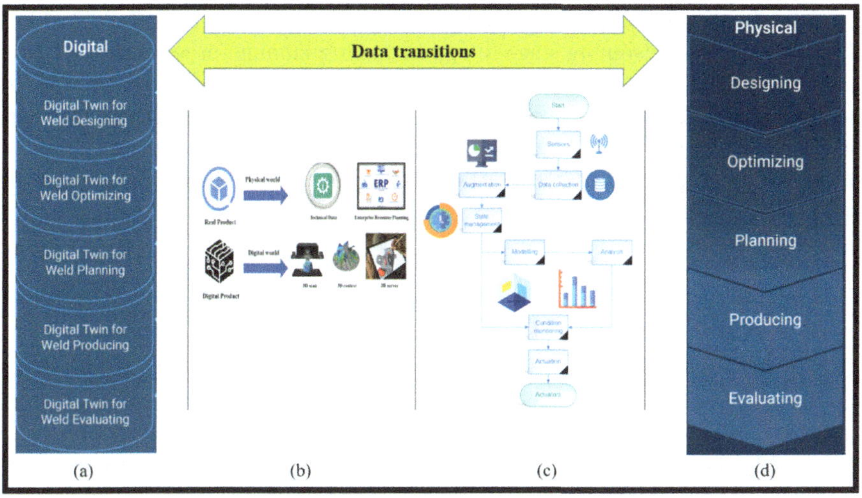

Fig. 4 Digital twin model. **a** Digital stage, **b** simulation and inspection, **c** monitoring stage, **d** physical stage

and presents personalized services. From a lifecycle view, the intelligent manufacturing system is classified into three aspects: framework, enabling technology, and sustainable, smart manufacturing.

The enabling technology of sustainable and intelligent manufacturing consists of digital twin-based, big-data-driven, artificial intelligence-driven, and Internet of Things (IoT)-driven. A twin model is proposed as the framework of digital twin-driven sustainable, intelligent manufacturing, especially for the laser welding process, Fig. 4.

Primarily, sustainable and intelligent digital twin-driven consists of basic sustainable and smart manufacturing principles and platforms. The data of the primary platforms comes from the equipment layer of the platform, which includes equipment, unit, production line, and production workshop. After the platform collects the data from the device layer, it combines cloud computing, AI, IoT, and other technologies to examine environmental, economic, and social factors comprehensively. It combines humans, equipment, and technology to provide virtual and physical prototyping data. Sensors may be included in the physical welding processes to verify the digital twin for improving numerical simulations' accuracy compared to the actual measurements in the physical welding tests. The role of digital twins in the optimization of the welding is obtaining the welding process window and assisting in optimizing and monitoring the biological welding processes. The integration of machine learning and AI into the digital twins can lead to the improvement in autonomous decision-making capabilities.

Digital twins will predict every possible root cause of the welding issues. Simultaneously, machine learning and AI will automatically recognize the issues and make decisions to solve the problems, correct errors, or indicate failures.

Interactions connect all the welding digital twins through the whole welding lifecycle, from the designing stage to optimization, planning, production, and evaluation. This makes the quality evaluation of the welded structures possible during the first stage of designing and through all other production phases. What can ensure the improvement in the welding process stability and the quality of the weld are the interactions and connections between the physical aspects of the welding with digital twins via AI and machine learning.

7 Sustainable Manufacturing and Laser Welding

Sustainability is the capacity to continue for a long period of time, considering three main objectives: economic considerations, social well-being, and environmental effects [22]. This structured approach provided an overall structure to segregate all indicators that can be used by the end-user to assess sustainability. A sustainable manufacturing process is one that realizes economic profit while minimizing negative environmental impacts having conserved energy and resources. In addition to enhancing employee and community safety, sustainable manufacturing improves product safety for a particular application. It is becoming increasingly important for industrial production to take a sustainable approach to resource and energy use because of new energy laws and eco-design directives, as well as cost-effectiveness and companies' environmental footprint.

A Life Cycle Assessment (LCA) is currently the best method to evaluate the environmental and social impacts of a process or product [23, 24]. As a proven methodology for evaluating environmental impact at the process or product level, as well as preventing responsibility shifting from one phase to another, it has been proven effective time and again. Social life cycle assessment of products (SLCA) is defined in the Guidelines for Social Life Cycle Assessment of Products (UNEP 2009) as a methodology that analyzes the possible positive and negative effects that products and services have on human beings throughout their lifetimes, such as health or wage issues. The demand is also increasing for products and goods made with sustainable methods, and customers are possible willing to pay a higher price for them. Manufacturing will continue to be a major influencer on economies, social issues, such as worker's compensation and Occupational Safety and Health (OH&S); and environmental issues, such as energy consumption, wastes, and emissions. Due to the linking of sustainable manufacturing to all or virtually all of the indicators of national and organizational sustainability, Sustainable Manufacturing (SM) emerged and is rapidly growing as a means to address "the sustainability challenge" that we are facing, through innovative systems, models, processes, and technologies. A batch of indicators is associated with each of these categories that would quantify the process's performance and production chain [25]. Actually, Industry 4.0 and sustainability are linked together in modern manufacturing and technology especially in the past decade. On the one hand, Industry 4.0 can help in achieving sustainable development, also assuring, for example, the preservation of resources on behalf of future generations. On the other, firms should be able to effectively use tools and opportunities pertaining to Industry 4.0 in shaping their organization, strategies, policies, and operations to achieve sustainable development and/or foster sustainability at a more general level [26].

A smart factory correlates with Industry 4, sustainable manufacturing, and the preceding. Sustainability can be viewed in different ways according to its purposes and applications, but the most widely accepted dimensions include the environment, society, economy, technology, and performance management. In the context of sustainable development, Environment, Society, and Economy are usually referred to as "Triple Bottom Line (TBL)". In order to achieve sustainability, manufacturing processes and products must be designed in a way that poses zero environmental impact and is 100% recyclable. Digital technologies help create products and processes, but to create sustainable development, they need to be converged with sustainability. In order to reap the benefits of Industry 4.0, which is sustainable manufacturing, manufacturers are currently focusing on such convergence. As part of the implementation of Industry 4.0 technologies, experts and researchers are addressing challenges and issues related to sustainable manufacturing through TBL. In Industries 4.0, the environment will be tackled in a number of ways, including climate change, resource depletion, and environmental protection. A new view has been brought to Industry 4.0, which is traditionally seen as a strategy to digitize operations and reap the rewards. In order to achieve this, Industry 4.0 technologies must be fully converged and coherent. Integration of all production areas, distributors, and customers is required to transform a manufacturing establishment into an intelligent factory. By providing seamless integration between Industry 4.0 technology platforms and information and communication technology platforms, Industry 4.0 technologies enable more transparency across production processes and supply chains, enabling better utilization of energy and resources. Every aspect of manufacturing is affected by such connected operations, producing massive amounts of data. All of these data contribute to the development of environmental, socioeconomic, and societal strategies when they are transformed into useful information [26, 27].

Through its streamlined manufacturing process and recycling and remanufacturing initiatives, Industry 4.0 contributes to the reduction of waste generation in sustainable manufacturing. By incorporating different types of sensors, for example, any manufacturing process or operation becomes considerably more transparent. In addition to collecting valuable information, such sensors also provide valuable information on behavior, usage, failure models, performance indicators, emissions, and performance under the stress of the product. Through the use of various simulation systems, such information is used to develop better products and processes while reducing environmental impacts without harming the company's competitiveness [27–29]. A similarly integrated system can also manage and monitor losses incurred in the manufacturing and use of a product during its life cycle. As a result, manufacturers can develop innovative products that are competitive yet environmentally friendly, thereby creating sustainable products. By allowing the development of equipment with much lower costs through the use of energy and resources efficiently through IoT, Artificial Intelligence, Machine Learning, Machine Vision, and Data

Analytics, Industry 4.0 offers a number of advantages from an economic perspective. In order to lower their operating costs, manufacturers are seeking ways to improve their operations [30]. Although manufacturers do well to reduce costs by developing strategies to reduce waste (arising from manufacturing and maintenance activities), decreasing productivity, and increasing energy consumption, challenges continue to hinder their efforts. With the implementation of Industry 4.0 technologies, manufacturers will be able to visualize how they are optimizing and non-optimizing their value chains. By taking advantage of such solutions, manufacturers can optimize their operations and reduce operational costs while increasing productivity by right-sizing their workers, facilities, and resources. Among other things, manufacturers will be able to reduce their waste generation through the use of modern and cleaner technologies for manufacturing and 3D printing. Industry 4.0 contributes to better products, which in turn benefit society as a whole in regards to the social dimension of sustainable manufacturing. In addition to this, better jobs will be created, which will result in a higher level of skill sets for workers. Consumers are expected to receive incentives from several manufactures to encourage them to return their old products and help with recycling and remanufacturing.

In modern manufacturing systems, laser welding is one of the most important elements of the manufacturing field. Following the selection of the welding processes to be assessed for laser welding applications, the relevant sustainability performance categories for welding need to be selected. Regarding the nature of the process, sustainability depends on a number of factors during the welding process. This includes welding speed, power, keyhole stability, automation, and control, as well as the fabrication of auxiliary materials. Energy is also used in the extraction of raw materials, the processing of raw materials for welding filler metals, and the manufacturing of auxiliary materials. Therefore, the production of the individual components of a welding system is as important as the procurement of resources [31].

An important element of welding training is the sequence and procedure, not only from a safety viewpoint, but also to improve production. Utilizing the appropriate resources can also make welder training more sustainable. Virtual reality offers novice welders a virtual training system, the virtual welding simulation system, by which they can learn basic skills and torch control. In addition to saving resources, virtual training or augmented reality reduces costs because consumables are not required. As a result, this type of training is very beneficial to sustainability. In welding technology, sustainability depends upon numerous factors, such as selecting the right welding process for the application, increasing welding speed, and reducing rework and rejection rates. Also, it is worth mentioning that it is necessary to select the right welding system before all these parameters [32, 33].

To survey the sustainability parameters of some welding processes and also their comparison, Table 2 has been presented. In this comparison study, nine welding processes, namely, shielded metal arc welding (SMAW), gas metal arc welding (GMAW), gas tungsten arc welding (GTAW), plasma arc welding (PAW), submerged arc welding (SAW), magnetic pulse welding (MPW), ultrasonic welding (USW) friction-based welding (FW), and laser welding (LW) have been included. The most important parameters of sustainability that have been studied include

Table 2 Comparison of sustainability factors for different welding processes

Weight: 1~3 Point: 1~5 Score=Point*Weight			Type of Welding Process																	
			1		2		3		4		5		6		7		8		9	
			Shielded Metal Arc Welding		Gas Metal Arc Welding		Gas Tungsten Arc Welding		Plasma Arc Welding		Submerged Arc Welding		Magnetic Pulse Welding		Ultrasonic Welding		Friction-based Welding		Laser Welding	
		Weight	Point	Score	Point	Score	Point	Score	Point	Score	Point	Score	Point	Score	Point	Score	Point	Score	Point	Score
Sustainability concerns	Energy Consumption	2	1	2	1	2	1	2	1	2	2	4	4	8	5	10	5	10	3	6
	Destruction of the Environment	3	1	3	2	6	2	6	2	6	4	12	4	12	4	12	5	15	3	9
	Welding Quality and Accuracy	2	2	4	3	6	3	6	3	6	4	8	4	8	5	10	4	8	5	10
	Reproducibility	2	2	4	3	6	3	6	3	6	4	8	5	10	5	10	5	10	5	10
	Process Stability	2	1	2	3	6	3	6	3	6	3	6	4	8	2	4	2	4	5	10
	Operationality	1	3	3	3	3	3	3	3	3	4	4	5	5	5	5	5	5	5	5
	Productivity and Efficiency	3	1	3	2	6	2	6	3	9	3	9	4	12	5	15	5	15	4	12
	Process Cost	2	5	10	4	8	4	8	3	6	2	4	2	4	3	6	2	4	2	4
	Applicability	3	2	6	3	9	3	9	3	9	4	12	4	12	3	9	4	12	5	15
	Process Complexity	1	5	5	4	4	4	4	4	4	2	2	2	2	2	2	2	2	3	3
	Process Limitations	2	2	4	2	4	2	4	2	4	2	4	3	6	4	8	4	8	5	10
	Adaptability	2	3	6	3	6	3	6	3	6	3	6	4	8	3	6	3	6	5	10
Total Score				52		66		66		67		79		95		97		99		104

energy consumption, destruction of the environment, welding quality and accuracy, reproducibility, stability, operationality, productivity and efficiency, cost, applicability, complexity, limitations, and adaptability. These parameters have been chosen according to the latest studies about the various welding processes and their sustainability [34, 35].

In this comparison process, a weight value is allocated to each parameter. The scores of each of the parameters assigned to each of the processes are calculated from the product of the considered points and the related weight. Finally, the final score of each process is obtained from the sum of the scores of all of the parameters.

As can be seen from Table 2, the total scores related to the sustainability parameters for the GMAW and GTAW processes are similar to each other. While the welding quality of GTAW is slightly better than GMAW, they are generally categorized at the same level of welding quality and accuracy. The total score of the PAW process is slightly more than processes GMAW and GTAW, which is due to the relatively higher efficiency of the mentioned process.

The SAW process has achieved a higher score in most parameters compared with the previous processes, which in total has caused a significant score difference for this process. The total scores related to the sustainability parameters of the MPW, USW, and FW processes are close to each other. In fact, the mentioned processes are generally at the same level in this regard, but the scopes of their applications are varied from each other.

Comparing the advantages of different processes, this can be concluded that laser welding is at a higher level than that of the MPW, USW, and FW processes. The mentioned point is due to the significant advantages of the LW method compared to other methods, in most parameters, including welding quality and accuracy, stability, productivity and efficiency, applicability, complexity, limitations, and adaptability.

8 Conclusion and Future Research Opportunities

Although laser welding is one of the most preferred fabrication methods, there are still some difficulties and challenges in employing this welding process and should be considered when integrating Industry 4.0. The HAZ (heat-affected zone), loss of alloying elements, porosity, and other defects such as cracks and insufficient penetration welds affect the resultant mechanical properties like the formability of the welded structure and an acceptable combination of input variables can control these failures. The challenge of obtaining the optimum combination of variables through a large number of experiments and a huge volume of data and deciding on which types of data to be analyzed should be addressed. Another challenge of the digitalization of laser welding is online monitoring. For this purpose, programming skills, knowledge of different software, and a secured transferring network to receive data from sensors and change them into readable data are needed [36]. Regarding the fact that enhanced welding quality and efficiency is a crucial part of intelligent manufacturing, in this chapter, the expansion of smart sensors, latest gadgets, and AI-based ways of real-time inspection and monitoring of welding quality are reviewed in detail:

1. In-process welding inspection is a perfect real-time monitoring procedure since the data obtained through it can be applied for adjusting welding conditions in real time.
2. Inspection monitoring devices are classified and reviewed.
3. Smart techniques and sustainable structure in welding are reviewed.

AI has a high potential to process and mine the data and is an asset in achieving multiple monitoring objectives. Therefore, a developed smart quality evaluation system becomes the most impressive and challenging one.

Toward future research, smart monitoring will concentrate on three features:

- An innovative acquisition platform of various welding signals.
- In-depth study of signs.
- Feedback control of welding variables.

However, to be more precise, the entire process of welding inspection and monitoring remains humanized. With regard to the various welding methods and their applications in repairing, fabrication of medical gadgets, computer devices and components, micro-welding, automotive, aerospace, and electronic industry, seam welding is a novel topic in this concept which is considered in modern manufacturing.

References

1. Pivoto DGS, de Almeida LFF, da Rosa Righi, Rodrigues JJPC, Lugli AB, Alberti AM (2021) Cyber-physical systems architectures for industrial internet of things applications in Industry 4.0: a literature review. J Manuf Syst 58:176–192

2. Nielsen P, Holm JR, Lorenz E (2021) Work policy and automation in the fourth industrial revolution. Glob New Emerg Technol Sustain Dev, Routledge, p 189–207
3. Santos JL (2021) Optical Sensors for Industry 4.0. IEEE J Sel Top Quantum Electron 27:1–11
4. Aminzadeh A, Parvizi A, Safdarian R, Rahmatabadi D (2020) Comparison between laser beam and gas tungsten arc tailored welded blanks via deep drawing. Proc Inst Mech Eng Part B J Eng Manuf. https://doi.org/10.1177/0954405420962391
5. Aminzadeh A, Parvizi A, Moradi M (2020) Multi-objective topology optimization of deep drawing dissimilar tailor laser welded blanks; experimental and finite element investigation. Opt Laser Technol 125. https://doi.org/10.1016/j.optlastec.2019.106029
6. Aminzadeh A, Karganroudi SS, Barka N (2021) A novel approach of residual stress prediction in ST-14/ST-44 laser welded blanks; mechanical characterization and experimental validation. Mater Lett 285:129193
7. Aminzadeh A, Karganroudi SS, Barka N, El Ouafi A (2021) A real-time 3D scanning of aluminum 5052–H32 laser welded blanks; geometrical and welding characterization. Mater Lett 296:129883
8. Velmurugan RS, Dhingra T (2021) Intelligent Predictive Maintenance: Industry 4.0. Asset Maint Manag Ind, Springer, 113–35
9. Wang H, Liu J, Xu Y, Wang X, Ren N, Ren X et al (2021) Experimental characterization and real-time monitoring for laser percussion drilling in titanium alloy using transverse electric field assistance and/or lateral air blowing. J Manuf Process 62:845–858
10. Wang L, Liu W, Chen Y, Chen H et al (2021) In-Process Visual Monitoring of Penetration State in Nuclear Steel Pipe Welding. Springer, Trans Intell Weld Manuf, pp 193–200
11. Antony J, Schlather D, Safronov G, Schmitz M, Van Laerhoven P, others. Toward fault detection in industrial welding processes with deep learning and data augmentation. ArXiv Prepr ArXiv210610160 2021
12. Stavridis J, Papacharalampopoulos A, Stavropoulos P (2018) Quality assessment in laser welding: a critical review. Int J Adv Manuf Technol 94:1825–1847. https://doi.org/10.1007/s00170-017-0461-4
13. Taneja NK (2021) Airlines in a post-pandemic world: preparing for constant turbulence ahead. Routledge
14. Xu Y, Wang Z (2021) Visual sensing technologies in robotic welding: recent research developments and future interests. Sensors Actuators A Phys, 112551
15. Xiao X, Liu X, Cheng M, Song L (2020) Towards monitoring laser welding process via a coaxial pyrometer. J Mater Process Technol 277:116409
16. Gao JQ, Qin GL, Yang JL, He JG, Zhang T, Wu CS (2011) Image processing of weld pool and keyhole in Nd:YAG laser welding of stainless steel based on visual sensing. Trans Nonferrous Met Soc China 21:423–8. https://doi.org/10.1016/S1003-6326(11)60731-0
17. Cai W, Wang JZ, Jiang P, Cao LC, Mi GY, Zhou Q (2020) Application of sensing techniques and artificial intelligence-based methods to laser welding real-time monitoring: a critical review of recent literature. J Manuf Syst 57:1–18. https://doi.org/10.1016/j.jmsy.2020.07.021
18. Sebestova H, Chmelickova H, Nozka L, Moudry J (2012) Non-destructive real time monitoring of the laser welding process. J Mater Eng Perform 21:764–9
19. Li L, Brookfield DJ, Steen WM (1996) Plasma charge sensor for in-process, non-contact monitoring of the laser welding process. Meas Sci Technol 7:615
20. Goel R, Gupta P (2020) Robotics and Industry 4.0. Adv Sci Technol Innov, 157–69. https://doi.org/10.1007/978-3-030-14544-6_9
21. Kumar A, Nayyar A (2020) A Roadmap to Industry 4.0: Smart Production, Sharp Business and Sustainable Development. Adv Sci Technol Innov 187–205
22. Schneider D, Paul M, Vernim S, Zaeh MF (2021) Sustainability assessment of manufacturing systems—a review-based systematisation. Towar Sustain Cust Bridg Smart Prod Manuf Syst, 1023–1030
23. Daniyan I, Mpofu K, Ramatsetse B, Muvunzi R (2021) Comparative analysis of the life cycle assessment for the assembly operations of railcar components. Procedia CIRP 104:1632–1637

24. Bevilacqua M, Ciarapica FE, D'Orazio A, Forcellese A, Simoncini M (2017) Sustainability analysis of friction stir welding of AA5754 sheets. Procedia CIRP 62:529–534
25. Seliger G (2012) Sustainable manufacturing for global value creation. Sustain Manuf, Springer; p 3–8
26. Favi C, Campi F, Germani M (2019) Comparative life cycle assessment of metal arc welding technologies by using engineering design documentation. Int J Life Cycle Assess 24:2140–172
27. Mashhadi AR, Behdad S (2018) Ubiquitous life cycle assessment (U-LCA): a proposed concept for environmental and social impact assessment of industry 4.0. Manuf Lett 15:93–96
28. Mehta KP (2019) Sustainability in welding and processing. Innov Manuf Sustain, Springer; 125–45
29. Ferrari AM, Volpi L, Settembre-Blundo D, García-Muiña FE (2021) Dynamic life cycle assessment (LCA) integrating life cycle inventory (LCI) and Enterprise resource planning (ERP) in an industry 4.0 environment. J Clean Prod 286:125314
30. Lee J, Davari H, Singh J, Pandhare V (2018) Industrial Artificial Intelligence for industry 4.0-based manufacturing systems. Manuf Lett 18:20–3
31. Baba ZU, Shafi WK, Haq MIU, Raina A (2019) Towards sustainable automobiles-advancements and challenges. Prog Ind Ecol an Int J 13:315–31
32. Sharma A (2018) A fundamental study on qualitatively viable sustainable welding process maps. J Manuf Syst 46:221–230
33. Saad MH, Darras BM, Nazzal MA (2021) Evaluation of welding processes based on multi-dimensional sustainability assessment model. Int J Precis Eng Manuf Technol 8:57–75
34. Vimal KEK, Vinodh S, Raja A (2015) Modelling, assessment and deployment of strategies for ensuring sustainable shielded metal arc welding process—a case study. J Clean Prod 93:364–377
35. Fotovvati B, Wayne SF, Lewis G, Asadi E (2018) A review on melt-pool characteristics in laser welding of metals. Adv Mater Sci Eng 2018
36. Zhao H, White DR, Debroy T (1999) Current issues and problems in laser welding of automotive aluminium alloys. Int Mater Rev 44:238–266. https://doi.org/10.1179/095066099101528298

The Role of Additive Manufacturing in the Age of Sustainable Manufacturing 4.0

Mina Ahmadi, Davood Rahmatabadi, Armin Karimi, Majid Haji Ali Koohpayeh, and Ramin Hashemi

1 Introduction

Industrial revolutions throughout history have been defined by the characteristics of the different emerging technologies of each moment. These features and the latest technologies develop the industry's methods of production at elevated speed. At the same time, they stimulate economic and social change, thoroughly transforming humanity and how it has grown and evolved.

Society has been into four industrial and technological revolutions, starting from the nineteenth century, which led to progress in all its aspects, both economically and socially. The First Industrial Revolution took place in the eighteenth century, aimed at reducing human efforts and upgrading manual production with the help of steam-powered engines. The second Industrial Revolution has mainly focused on producing standard parts to raise the level of production speed, though design and flexibility were not high. The third revolution, with the aim of quality, flexibility, and speed augmentation, has brought modern technologies and automation into manufacturing [1].

M. Ahmadi
School of Automotive Engineering, Iran University of Science and Technolgy, Tehran, Iran

D. Rahmatabadi (✉)
School of Mechanical Engineering, University of Tehran, Tehran, Iran
e-mail: davood.rahmatabadi@yahoo.com

A. Karimi
Department of Aerospace Engineering, Sharif University of Technology, Tehran, Iran

M. H. A. Koohpayeh
Mechanical Engineering Faculty, K. N. Toosi University of Technology, Tehran, Iran

R. Hashemi
School of Mechanical Engineering, Iran University of Science and Technology, Tehran, Iran

H. Gholami et al. (eds.), *Sustainable Manufacturing in Industry 4.0*,
https://doi.org/10.1007/978-981-19-7218-8_4

With the advent of intelligent automation technology, following these three phases of revolution is The Fourth Industrial Revolution, also known as Industry 4.0. Implementing modern manufacturing systems and integrating information technologies is paramount for manufacturers in Industry 4.0 to stay economically competitive. Industry 4.0 tends to provide a more intertwined, flexible, and productive manufacturing process and supply chain, producing customized products based on each customer's needs with less possible delivery time than the mass production system. By reconsideration of human roles and the digital connection of production systems, Industry 4.0 proposes a cyber-physical approach to initiate such intelligent factories [2].

The main pillars of Industry 4.0 in automation are Internet of Things (IoT), Big Data, and Cloud Computing, along with physical aspects of advanced manufacturing technologies, including Robots and Additive Manufacturing (AM) [1, 2]. The decision-making process becomes available by collecting and analyzing a great deal of statistical data (cloud computing and big data analytics) from physical devices through IoT. IoT merges the real world with the virtual world to minimize human intervention and improve efficiency and accuracy [1, 2]. Such records, together with data from design, delivery, and logistics, are Big Data that can empower organizations to achieve production excellence by meeting customers' changing needs in the shortest possible time, reducing costs, and customizing freely designed products [3].

Accessibility to shared location-independent data is another primary concept of Industry 4.0 provided by cloud computing. Cloud computing can reduce the total number of interactions by making accessible virtual communication between operators and machines. It can also analyze data from market and customer points of view and provides all suggestions for individuals, equipment, and procedures automatically. Remote controlling and investing in CNC machines and robots, material flow tracking through the manufacturing process, extracting features of parts, and using Big Data to optimize future parts are clear manifestations of using cloud computing in manufacturing. These cyber technologies ensure the emergence of smart factories [4].

When it comes to the physical aspects of Industry 4.0, robots and additive manufacturing play crucial roles. These technologies, with the help of cyber systems, improve productivity at a lower cost. Robots can carry out tasks autonomously in any environment, improving manufacturing performance, warehousing, and monitoring [1].

Due to the mass customization nature of Industry 4.0, areas of manufacturing need to be reshaped to meet this demand, as traditional manufacturing methods limit the capability of factories [2]. Additive manufacturing, mainly defined as a process by which sophisticated solid objects are made layer by layer, is critical in fabricating customized parts with advanced materials and complex geometries [2, 4]. AM is widely used in a variety of industries, from biomedical to aerospace and engineering [4]. In comparison with traditional manufacturing methods, AM can empower all areas of Industry 4.0 by streamlining the manufacturing processes [5]. All products can be printed at a specific lower speed, and as a result, logistic costs would be reduced [1].

There are also specific features that can identify materials and machines for a given part. By evaluating the manufacturing process, storing designing and machining rules, and considering customer use, resources can be automatic scaling. One of the main limitations of conventional methods is tooling supply and manufacturing correct tooling, which can be inhibited through AM due to its tool-free nature. AM can also play the role of "rapid tooling" for the conventional approach as needed [1]. A distinct advantage of AM is introducing intelligent materials to the industries that can, in turn, provide reconfiguration possibility of printed parts and obtaining desired mechanical properties [2].

Considering electrical energy consumption and material consumption, Industry 4.0 aims to implement sustainable manufacturing methods. In other words, its goal is to minimize waste and diminish environmental impacts. AM can reduce energy and material consumption rates by using the concept of a layer-by-layer production method and optimizing manufacturing orientation. To this should be added that AM is more cost-effective and energy-efficient when it comes to producing plastic parts than traditional methods [6].

Adopting (AM) and other advanced manufacturing technologies seem to herald a future in which value chains are faster, smaller, more localized, more cooperative, and offer significant sustainability benefits. However, despite these future benefits, AM has not been sufficiently explored from a sustainability perspective and its contribution to Industry 4.0.

Therefore, this chapter focuses on the various additive manufacturing methods and their relationship with different Industry 4.0 components. Finally, the topic of AM through the lens of industrial sustainability is explored to provide a more comprehensive understanding of the implications of AM for improving the sustainability of industrial systems.

This chapter begins by providing.

- An overview of AM,
- Materials used in AM, and
- AM different processes.

and continues by giving information on.

- The interrelationship between additive manufacturing and Industry 4.0 components, such as the Internet of Things, Big Data, Cloud Computing, and Robots.

and finishes by describing the ways in which AM can enable more sustainable models of production and consumption and the challenges of implementation of AM in an extended application.

2 Additive Manufacturing Materials

Industry 4.0 is moving in the field of intelligent technology and automation. In recent years, combining new and modern technologies such as additive manufacturing with information technology for economic competition as a new field in the manufacturing sector has received more attention. Elimination of geometric constraints for designers and manufacturers, reduction of waste materials, production and assembly stages, and control of consumables are the main advantages of 3D printing methods. These benefits reduce costs and production time. In 3D printing processes, potential expenses include the cost of machinery, human resources, energy consumption, and raw material, eliminating the costs of mold preparation and integrated production reduces production costs. With the development of 3D printing methods, utilizing different materials and enhancing consumables are getting more attention [7]. Today, various 3D printing methods can use almost all materials in the industry, and advances are obtained in printing intelligent materials, ceramics, metal alloys, concrete, polymer-based composite reinforced with particles, and continuous fibers. In other words, 3D printing methods cover relatively all materials in different industries. Limitations have been solved, and new 3D printing fields, such as 4D printing, have been achieved. In the last two decades, smart material printing or so-called four-dimensional printing has become one of the most attractive topics in the field of additive manufacturing. In fact, by designing innovative materials and adding printing capabilities to them, 4D printing can be used without additional equipment compared to 3D printing [8, 9].

Innovative smart materials are a particular group of materials that can stabilize the deformation and different shapes. If proper stimulation is applied, those materials will be able to return to their original form. In fact, by using and printing these materials in different methods of 3D printing, the time dimension is also added, and the printed structures can deform over time, which is possible by applying different stimuli. In general, shape memory materials (SMM) fall into three categories polymers, alloys, and ceramics, of which polymers and alloys are the most popular. The limitations of ceramic printing due to the need for equipment with high working temperature capability have led to the fact that the 3D printing method has not been used to make ceramic structures. In the following, the two common categories of alloy and polymer, formal memory, and the mechanisms of each are briefly discussed.

2.1 Shape Memory Alloys

Shape memory alloys (SMA) are a new group of smart materials that, if subjected to appropriate heat treatment with a specific chemical composition, will show the ability to return to a predetermined shape or size. It means SMAs can recover to their original shape if heated to a specific temperature. These materials are also capable of converting heat (electrical) energy into mechanical energy. If the heating and cooling of these alloys are controlled by electric current, repeatable cyclic shapes and movements would be created several times in a row. SMAs have two unique characteristics: Shape Memory Effect (SME) and Pseudo elastic behavior [10]. Other features of these alloys are

- High corrosion resistance.
- Relatively high electrical resistivity.
- Fairly good mechanical properties.
- Long fatigue.
- High elasticity.
- Adaptability to the body.

The most important application of these alloys is in the aerospace and medical industries. These alloys in most cases include Ni–Ti, Cu–Zn-Al, and Cu-Al-Ni [10].

The primary mechanism that controls the properties of SMAs is the crystal changing of the alloy. It means that the martensitic structure becomes an austenitic structure at high temperatures, and during the cooling process, the reverse of this process occurs. Many materials have martensitic transformations, but the distinct advantage of SMAs over other alloys is the twining phenomenon in the martensitic phase. While other materials are deformed by slippage and dislocation movement, SMAs react to stresses by changing the simple direction of their crystal structure through the twin boundaries [10]. Suppose a plastic deformation occurs in these alloys at low temperatures where the martensitic phase is predominant. In that case, a twin crystalline structure is formed for the alloy due to the plastic deformation. The original shape can be restored by heating the deformed alloy to the starting temperature of the austenite phase. This ability is called the Shape Memory Effect (SME) and results from the changes of the martensitic phase at low temperature to the austenite phase at high temperature [10]. In the shape memory phenomenon, the sample is deformed to a certain amount in a completely martensitic state. By heating the sample and returning it to the austenitic state, the shape of the sample returns to its original condition.

2.2 *Shape Memory Polymers*

Although Shape Memory Polymers (SMPs) were introduced decades later, compared to SMAs, they have some advantages, including

- More straightforward processing.
- Lower cost.
- Lower density.
- Higher flexibility.
- Higher formability.
- The ability to manipulate mechanical properties.
- Shape memory stimulation with various stimuli.

For this purpose, they are getting more attention. As mentioned in the previous section, the SME has two main features: stabilizing the shape and recovering the original shape. The ability to stabilize is the ability of the memory material to change from the original shape to the temporary shape through the programming process. Recoverability also indicates the power of the material to recover the original form. In the programming process, the SMP is mechanically deformed, and the modified shape is temporarily stabilized. The most important feature of the memory effect is stabilizing this quick and deformed form, which should not be changed by removing the stimulus. The shape memory cycle for polymers consists of three stages: programming, storage/shape stabilization, and recovery [11]. Thermal energy is applied in the programming stage as long as the temperature is above the material transfer temperature. At this stage, it is possible to achieve the desired shape by applying force because it is very soft and rubbery. The storage phase requires the maintenance of external load and pressure, and the part is placed into the desired shape temporarily. Consequently, the temperature reduces below the transfer temperature. In shape memory polymers, two parts are placed together as the soft part and the hard part. One is responsible for remembering the original shape, and the other is responsible for energy storage. In other words, for an SMP to be able to exhibit this property, it must have a force-storage phase and a force-retaining phase, by which external stimulation releases the stored force and the matter returns to its original shape. At a glance, SMPs can be divided into three categories based on their chemical structure, types of stimuli, and shape memory function.

Of course, it should be noted that these SMPs have weaknesses compared to SMAs, the most important of which being low thermal and electrical conductivity and low mechanical properties, which leads to reduced recovery speed and recovery force. This leads to the emergence and widespread use of various reinforcements to expand polymer composites to overcome the mentioned limitations.

2.3 Other AM Materials

In addition to shape memory materials (SMMs), high entropy alloys (HEAs), piezoelectric materials, conductive polymer, composite, multi-materials, etc. are materials with special properties that can be fabricated with different additive manufacturing techniques. One of the unusual properties of some ceramics and polymers is the piezoelectric effect. The piezoelectric effect is seen in many materials, including mono-crystals, ceramics, polymers, and composites. The generation of the electric potential difference in some non-conducting crystals, such as quartz, is all under reverse tension or pressure, and the higher the pressure or tension, the greater the potential difference produced [12]. By applying an external force, the dipoles of these ceramics are excited, and an electric field is created. Reversing the effect of force (for example, from tensile to compressive) changes the direction of the area. Piezoelectric materials are used in converters and devices that convert electrical energy into mechanical energy or vice versa. These materials are most commonly utilized in sensors, and piezoelectric sensors are mainly used in high-frequency sounds in ultrasonic transducers for medical imaging. Alloys have long been used to improve the properties of materials. In high entropy alloys (HEA), there are at least five essential elements with approximately equal atomic percentages. The important properties of these elements have attracted the attention of researchers over the past few years and have made rapid progress in various fields of research and application [13, 14].

3 Advanced Additive Manufacturing Processes

Charles Hall developed the first AM process in 1986, known as stereolithography, followed by further advances. 3D printing, which involves various methods, materials, and equipment, has evolved over the years and can transform production and assembly processes. Additive production has been widely used in multiple industries such as construction, prototyping, and biomechanics [15]. This technology allows researchers to create complex shapes that were previously impossible using traditional construction methods [16, 17]. By using 3D printing, researchers can create complex designs inspired by nature and multi-material designs [18, 19], remotely control robots [20], designs produced with machine learning and optimization algorithms [21, 22], drug delivery systems [9, 23], and even small environments for biological tissues [24, 25]. Common 3D printing technologies include material extrusion, photopolymerization process, powder bed fusion, material jetting, lamination process, and direct energy deposition [15]. Additive production methods have evolved to satisfy the printing demand of complicated components with good resolution. As an AM technology, rapid prototyping can print and fabricate large structures, reduce fabrication defects, and enhance mechanical properties, which are some of the main factors in developing AM technologies. The most commonly used AM method, which mainly utilizes polymer filaments, is fused deposition modeling. Production of powder additives by Selective Laser Sintering (SLS), Selective Laser Melting (SLM), inkjet printing technology, contour fabrication, stereolithography or (SLA), Direct Energy Deposition (DED), and Laminated Object Manufacturing (LOM) are examples of AM main methods. Their various applications, the appropriate materials for each technique, and their advantages and disadvantages are briefly described and introduced in the following sections. Figure 1 shows a schematic of four widely used additive generation methods.

3.1 FDM

In the Fused Deposition Modeling (FDM) method, the primary material used for 3D printed layers is a continuous filament or a string composed of thermoplastic polymer [27]. The string is heated in a nozzle, and its state changes to a semi-liquid

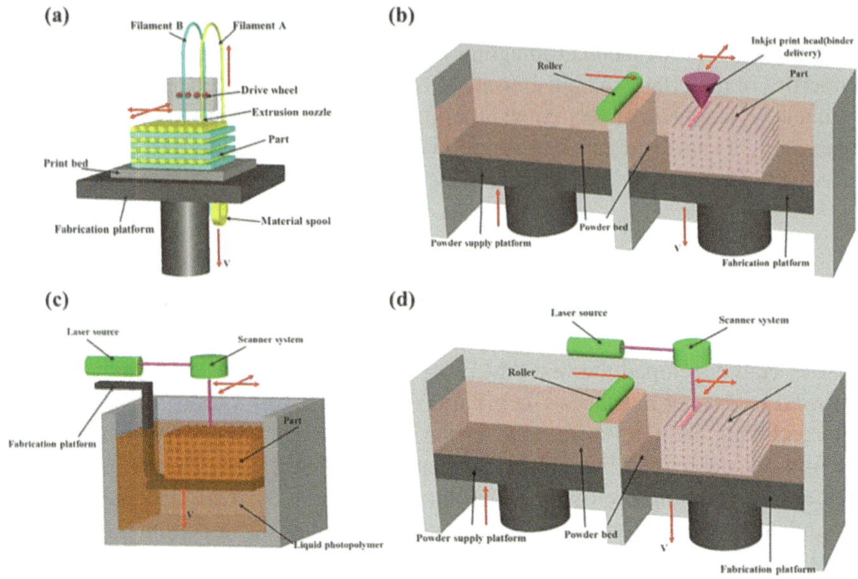

Fig. 1 Schematic of the main AM methods: **a** FDM, **b** inkjet printing; **c** stereolithography, and **d** SLM [15, 26]

and later extruded onto a print screen or previously printed layers. The thermoplasticity feature and rheology of the polymer filament are the main highlights of this method, allowing filaments to bond to each other during the printing process and then solidify at ambient temperature after the printing. Layer thickness, printing speed and temperature, raster width and direction, and the amount of free space (space between layers) are the central AM processing parameters that directly or indirectly have effects on the mechanical and microstructural properties of printed parts [27–29]. The distortion between layers is one of the leading causes of mechanical weakness in this method [30]. The main advantages of FDM over other AM methods are higher speed of printing, simplicity, and lower cost. However, weaker mechanical properties and lower quality of the finishing surface and layer-by-layer appearance [31], and limitations on the number and types of thermoplastic materials used by FDM are this method's drawbacks [32]. Developments in fiber/powder reinforced composites helped FDM enhance the properties of 3D printed components [33]. Still, fiber orientation, particle distribution, filler-matrix bonding, and cavities are principal challenges in printing 3D composite components [26, 33].

3.2 *Powder Bed Fusion*

In these processes, thin layers of very fine powders are spread and pressed on a plate—a laser beam bonds selectively powders of each layer in the presence or absence of a binder. Subsequently, layers of powders are then spread on the previous layers' surface, and by repeating the same method, a final 3D shape is produced. Excess powder is just then removed after the process has finished. If necessary, a finishing process and further details such as coating, baking, and penetration are used to improve the properties. The distribution of the powder and its compaction, which determine and affect the printed part's density, are the fundamental factors in this method [34]. Using laser is limited to powders, whose melting/sintering temperatures are low; otherwise, the laser is replaced by a liquid glue. While SLM's utilization of materials is limited to certain metals, including steel and aluminum and some of their alloys, SLS uses various polymers, metals, and alloy powders. Laser scanning's purpose during the SLS method is not to melt the powders entirely. But the increased local temperature of the grains' surfaces leads to the particle's fusion at the molecular level, while, in SLM, powders are completely melted and combined after laser scanning, which results in higher mechanical properties [35]. Regarding the use of liquid adhesive, which is also known as one of the 3D printing methods, the bond's rheology and chemistry, shape, and size of the powder particles, as well as the deposition rate, the interactions between powder and adhesive, and the post-processing techniques play an essential role in production parts in this method [26, 34]. The liquid adhesive's printed parts' porosity is frequently higher than those created by SLM or SLS [34]. The power and scanning speed of the laser are the main parameters that affect such manufacturing processes. More details and information about various types of lasers as well as their impacts on 3D printing processes can be found in the article by Lee et al. [35]. Higher quality and resolution of the printed structures are among the advantages of powder bed melting processes, making them appropriate methods for creating complex structures.

3.3 *Inkjet Printing and Contour Crafting*

Inkjet printing produces ceramic structures and prints complicated and advanced ceramics for scaffolding in tissue engineering applications. A stable suspension of ceramic, such as zirconium-oxide powder, is drawn into the water through an injection nozzle on the substrate and layered in drops [36]. These droplets then can create continuous patterns and, as they solidify, obtain sufficient strength for holding subsequent layers of the printed object. This method is fast, flexible, and efficient in designing/printing complex components. Two principal types of ceramic inks are wax-based inks and liquid suspension systems. The former solidifies by melting and being placed on a bed, while the latter solidifies as the liquid evaporates. Ceramic particle size distribution, ink viscosity, solid content, extrusion rate, as well as nozzle size, and printing speed are the determinants that affect the quality of inkjet-printed parts [37]. Low resolution and lack of adhesion among layers are the significant shortcomings of this method.

Like inkjet printing, the contour crafting process is the primary additive manufacturing method for large building structures. This method can extrude and eject concrete or clay paste with the help of larger nozzles at higher pressure.

3.4 SLA

SLA is among the pioneer additive manufacturing methods, which was developed in 1986. This method employs ultraviolet light (or electron beam) for initiating an interchain reaction of resins or monomer solution layers. Monomers (primarily based on acrylic/epoxy) can react to ultraviolet light and immediately turn into polymer chains after radicalization or, in other words, after activation. After polymerization, to keep the following layers, the desired pattern solidifies inside the resin layer. The nonreactive resin is also removed after printing. To obtain the determined mechanical characteristics, some printed parts may be supplemented by post-processing operations, like heating or optical functions. Ceramic particles' dispersion in monomers has the ability to be used for printing polymer-based composites [37].

3.5 DED

High-performance super alloys will be available through the direct energy deposition method. Laser-engineered lattice and solid laser formations, conductive light-fabrication, DMD or direct metal deposition, electron beam, and arc wire melting are other terms referring to DED [38]. DED uses an energy source (laser beam or electron) focusing directly on a small substrate area melting a raw material (powder or wire) simultaneously. The molten material then precipitates and solidifies after the laser beam moves [38]. The main difference between the two methods of DED and SLM is that there is no powder bed in DED, and the raw materials are fused layer-by-layer before deposition, similar to FDM. But much higher amounts of energy are needed to melt metals in DED than in FDM. So, it can help fill cracks and hardening of produced parts, which is limited in applying the powder bed fusion method. DED allows the simultaneous deposition of multiple axes and several materials [1]. Moreover, this technique can be combined easily with other conventional reduction processes like machining. This method usually uses titanium, Inconel, stainless steel, aluminum, and related alloys for aerospace applications. In general, the main characteristic of DED is high speed [39] and a vast operating range. However, it has a lower level of accuracy and quality and can produce parts with less complexity in comparison with SLS and SLM [38]. Therefore, this method is a commonly employed technique for huge components but with lower complexity. It is also an appropriate option for repairing large and difficult-to-repair parts. DED reduces fabrication time, needs lower investment, grants superior mechanical features, and controls the microstructure. It is used in repairing turbine engines and has different applications in various industries, including automotive and aerospace. Table 1 summarizes the materials, applications, advantages, disadvantages, and scope of separation of the primary additive production methods.

4 Interrelationship Between Additive Manufacturing and Industry 4.0 Components

Among all transformative technologies of Industry 4.0, AM is the only one that associates with manufacturing, an umbrella term that describes techniques with the capability of fabricating 3D objects layer upon layer. It offers many considerable advantages over other traditional manufacturing methods, and enabling the production of complex and challenging geometries might be the most notable one. Customization/personalization is another benefit of AM that turns it to be key manufacturing technology for Industry 4.0. However, AM and other Industry 4.0 elements can benefit from each other and connect the physical world of AM to the digitalized world of Industry 4.0.

4.1 The Relationship Between IoT and AM

Internet of Things which is known as the Industrial Internet of Things (IIoT) in the manufacturing world is a combination of devices and physical objects implanted with electronic gadgets, sensors, and software aiming at initiating the exchange, data, and information collection as well as facilitating communications of people, products, and machines. With the help of sensors attached to different machines and other physical objects during the fabrication stage and collecting data, IIoT affects real-time decision-making and leads to increased productivity and efficiency. Electronic objects used for fully functional IIoT were limited due to their production and implementation costs. However, due to being cost-effective and the ability of printing complex and functional electronic tools, AM came to help IIoT. Therefore, integrating AM and IoT/IIoT is essential for obtaining improved products and processes. As far as some factors such as risk, cost, and time are concerned, AM helps enhance manufacturing techniques for those electronic devices that their productions were impossible by subtractive strategies. For example, AM can remove many conventional constraints in producing sensor structures, such as those associated with planar electrical systems. The result is the possibility of creating thoroughly altered surface topographies that can promote the angled arrangement for sensor divisions [41].

Table 1 Summaries of the comparison of the main additive manufacturing methods [15, 26, 38, 40]

Process	Material	Applications	Advantages	Disadvantages	Quality
Filament Deposition Modeling	Filament, continuous thermoplastic and reinforced composite, and waxes	Prototypes, developed composite structures, tooling, casting patterns	Low cost, high speed, simpler process	Weak mechanical properties, limitation in used materials, layer structure	50–200 μm
Powder Bed Fusion	Compressed metals particles, (limited types of alloys and polymers for SLS and SLM, ceramic powder for 3DP)	Medical, aerospace, electronics, tooling, casting patterns, functional parts, heat exchanger, net shape lightweight components	High quality and resolution	Low speed, expensive, high porosity for 3DP	80–250 μm
Inkjet printing and contour crafting	UV curable acrylic plastic, wax, soil, and concrete	Medical, building large components	High speed, capability of printing large structures	No adhesion between the layers, layer structure	Inkjet: 5–200 μm Contour crafting: 25–40 mm
SLA	Photopolymer, liquid photosensitive resin curable with UV light	Medical, rapid prototyping	High quality and resolution	Limited materials, low speed, expensive	10 μm
DED	Metals and alloys in form of powder or wire, polymers, and ceramics	Medical, coating, aerospace, repairing, reinforcing	Lower cost and production time, improved mechanical properties, controllable microstructure	Lower accuracy, poor finishing surface, limitations in printing complex structures	250 μm
LOM	Paper, plastic, metal	Prototypes, casting models, electronics, intelligent structures, paper	Higher range of materials, suitable for constructing large structures, lower cost and production time	Lower accuracy, poor finishing surface, limitations in printing complex structures	Varies according to thickness

On the other hand, IIoT can help AM in different areas, such as monitoring customer interaction during the early stages. IoT enables a cloud platform for users to control and monitor the production process distantly. It is a platform which integrates 3D systems, materials, knowledge, and test data for printing, designing, and process planning. In other words, IoT is an excellent asset to AM optimization by collecting data from various sensors in real time and processing them by sophisticated digital techniques. It is now available for even mobile devices to use Wi-Fi and cloud platforms for online monitoring of 3D printers [2, 4, 41, 42].

4.2 The Role of Big Data Analytics in Additive Manufacturing

Based on the previous description of IIoT, it is a system of interconnected computing devices, and digital and mechanical machines with the ability to transfer data and information over a network without human-to-human or human-to-computer interaction. It means a large amount of data is gathered from different AM machines and needs to be processed quickly, letting users make informed decisions. Based on the previous description of IIoT, it is a system of interconnected computing devices, and digital and mechanical machines capable of transferring data and information over a network without any need for human–human or human–computer interaction. It means a large amount of data is gathered from different AM machines and needs to be processed quickly, letting users make their own decisions. This is where BDA comes into play. A collection of massive datasets that traditional databases and other software techniques cannot analyze or characterize big data. Like IIoT, BDA also plays a significant role in AM as it is capable of analyzing copious amounts of data. Due to rapid AM materials and systems developments, using big data for analysis is getting more critical. BDA comprises Company, Collaborators, Customers, Competitors, and Context (5Cs) in Industry 4.0 and cyber-physical[1] systems environment, which include intelligent cognition, configuration, cyber connection level, and conversation level of data to information. This enables a streamlined approach for analyzing data and, therefore, better product and system quality and reliability in the context of an Industry 4.0/intelligent industrial world of factories. Since AM involves complex and various interactions and connections between design, materials, fabrication processes, and part performance, enormous amounts of data need to be gathered and analyzed throughout the product life cycle for opportunities like cost and time reduction, defects detection, thermal distortion prediction, energy consumption optimization, and enhanced efficiencies. Figure 2 illustrates how big data helps AM to enhance the quality of products [2, 41, 43].

---

[1] Transformative technologies for managing interconnected systems between their physical assets and computational capabilities.

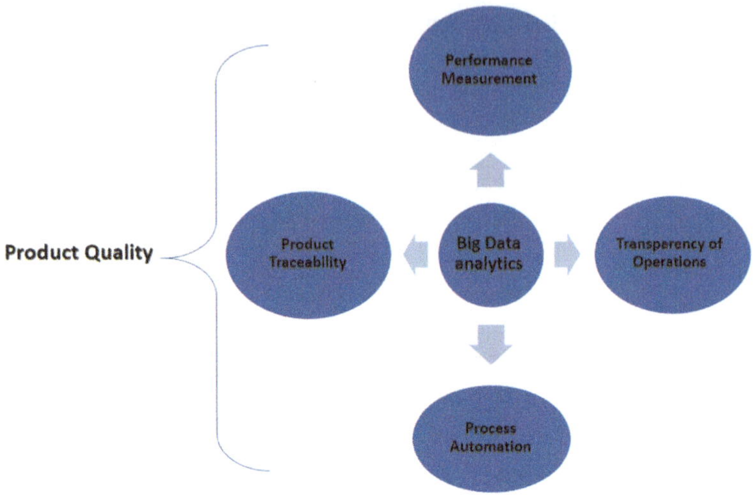

Fig. 2 The ways big data analytics are used to help AM

4.3 The Use of Cloud Computing in Additive Manufacturing

One major deterrent that organizations face is expensive software and hardware for analyzing a large amount of data and enabling on-demand, convenient access to a shared pool of computing resources, like servers, networks, etc. And this is the reason for Cloud Computing's (CC) popularity that without being bound to specific machines it can analyze data over the Internet. Now there is a rapid shift from IT resources in organizations to CC due to its offered benefits, such as high speed, productivity, and improved security and performance. CC goes hand in hand with IIoT, which enables collecting large amounts of datasets through connected devices. This data can be processed and analyzed with various strategies using CC aiming at cost and time-saving. Utilizing CC for AM is getting more attention as AM is capable of generating large datasets. One example is integrating sensors for data collection and then processing it via cloud-based systems and software, which results in AM optimization in terms of design, process, time, and cost [5, 41].

4.4 Industrial Autonomous Robots and Additive Manufacturing

An autonomous robot is defined by a certain degree of autonomy and self-sufficiency. It can perceive or is programmed to perceive its surroundings, settle on choices based on what it can see, and finally actuate movements/manipulation within the environment. Robots' roles in different industrial applications are significant; they are used in the medical, aerospace, automotive, and construction sectors. With increases in the quality and quantity of multipurpose robots, there are more developments in sophisticated robots. Additive manufacturing changes traditional manufacturing from assembly to logistics and robots, likewise, have profoundly impacted the industrial world. Combining these two (AM and robots) includes additive manufacturing for robot end effectors,[2] and robots used for 3D printing parts. The 3D printing robots can perform many tasks, ranging from new manufacturers to repairs to detecting damages and surveillance [2, 41].

5 Sustainable Manufacturing

Sustainability is any development in Reducing, Recovering, Recycling, Reusing, Redesigning, and Remanufacturing in three principal dimensions so that the present and the next generations can meet their needs. And manufacturing is defined as converting raw material into goods and other services. The efficiency of such a conversion process plays a crucial role in determining the environmental impacts associated with manufacturing [44].Green and sustainable manufacturing has appeared as a globally recognized mandate. The U.S. Department of Commerce defines sustainable manufacturing (SM) as "the processes of creating fabricated products that use non-polluting methods, conserve energy and natural resources, and are economically sound and safe for employees, communities, and society as a whole, and clients" [45].

In other words, sustainable manufacturing reduces the environmental impacts and enhances social and economic effects over the product's entire life cycle. SM promotes eco-efficient activities to minimize pollution and supports emerging sustainable innovations. Therefore, the integration of i4.0 components and principles for assessing and developing sustainable manufacturing can maximize the economic, environmental, and societal values of i4.0 [46].

Additive manufacturing—an emerging manufacturing process—not only has the potential to change the landscape for product development, manufacturing, and logistics, but it can also improve sustainability across a variety of industries. As a process in itself, AM already represents a more sustainable means of production than traditional methods. This is especially apparent due to the fact that 3D printing eliminates the usage of excess material and, therefore, unnecessary waste from the outset. Using generative design also plays a vital role in part optimization of AMed products and is one of the main benefits of 3D printing. In addition, a 3D printer allows on-demand manufacturing, which helps save time and eliminate long transport routes, and therefore reduces CO2 footprints.

The sustainability implications of adopting additive manufacturing can be classified into four stages of the product's life cycle:

- Designing the product and the process;

[2] In robotics, an end effector is a device at the end of a robotic arm designed to interact with the environment.

- Processing the input material;
- Make-to-order component and product manufacturing;
- Closing the loop.

In terms of product and process design, additive manufacturing makes the design and creation of more optimized and complex products achievable due to greater freedom in the product's shape and geometry, with fewer required assemblies and materials used. When it comes to component and product design and redesign, AM offers a variety of new design-free forms that allow easier fabrication and, therefore, easier maintenance of lightweight structures with higher operational efficiency, functionality, and durability.

Like improvements in product design, AM also makes improvements in the process design by offering more efficient energy and natural resource use processes.

As there is a combination of different AM technologies, so too is there considerable variation in the materials utilized as inputs, which can be processed via minimum resource use. Recycled materials can also be used as inputs for several AM processes, which is considered as a step toward the sustainable development of AM. Moreover, AM can also convert waste and by-products into products. Some studies demonstrate that materials traditionally classified as trash can be upcycled to fabricate luxury products utilizing AM.

AM enables making-to-order, customized, and personalized components at lower cost, while less waste is produced from a sustainability perspective in the economy. Attempts at closing the loop can also be obtained at different stages and scales in AM. Under a closed-loop system, businesses and companies reuse the same materials repeatedly to create new products. And these are how AM contributes to sustainable development [47].

In summary, AM has the potential of providing several sustainability advantages as listed here [4, 48]:

- Less raw material is required during the supply chain process.
- AM improves raw materials' efficiency (in powder, liquid, or wire form) through its feature called net shape manufacturing.
- In addition to its cost-efficiency and freedom of design, AM could become an energy-efficient and environmentally friendly manufacturing.
- Less waste material and therefore less pollution is produced.
- Products with higher efficiency and flexibility can be achieved by AM.
- It helps create parts/products for optimized performance, such as reduced weight, enhanced mechanical properties, and optimally designed components by incorporating gas flow paths and heating/cooling channels.
- It allows the production of customized elements in short batches, at the right moment, and according to customers' needs.
- The transportation costs decrease within the supply chains and transportation pollution.

Components produced by AM are lighter, need less raw materials, create less waste material during processing, and consume less energy. It also gives longer product

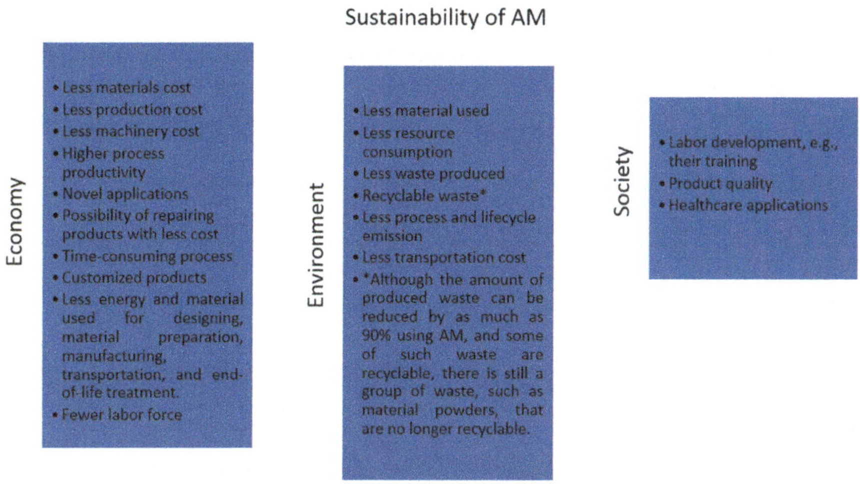

Fig. 3 Three aspects of AM's contribution to sustainability [49]

lifetimes by allowing replacement sections to be manufactured and remanufactured quickly enhancing the reliability and modular design that facilitates upgrading products. Therefore, this technology allows for shorter, simpler value chains by allowing versatility in place. Figure 3 summarizes three aspects of sustainability for AM.

6 Challenges, Drawbacks, and Limitations of AM

Additive manufacturing presents a great deal of opportunities for mass customization applications and moving toward Industry 4.0 needs. Theoretically suitable and more practical than traditional methods, there are some challenges and drawbacks for AM to overcome. High costs (for both machines and part production), limited material, defects, unsuitable finishing surface, little application (especially for large-scale parts), slow speed of printing, and anisotropic mechanical properties can be categorized as Additive Manufacturing drawbacks [15].

In comparison with traditional methods such as fusion modeling and casting, AM might be more time-consuming. In addition, layer-by-layer production with a high resolution imposes a high cost for materials and consumes a high amount of energy. Being expensive and time-consuming are the main limitations of AM in Industry 4.0, inhibiting mass production. Worthy of note also is that when a part is printed layer by layer, there would be inadvertent porosities affecting the part's mechanical properties as it reduces the interfacial bond of layers. Materials' microstructure can also vary inside each layer and boundary, resulting in anisotropic mechanical properties. This can emanate from either production methods or printing material [15].

Since AM is widely used in Industry 4.0, another major limitation of AM is that there are some restrictions on printing components' size. For instance, printing on-site buildings with the help of SLS or SLM systems is not applicable when it is too large. The size of the printing part affects the structural performance, and in fact, the dimension of the printed parts is determined based on the deposition approach. Besides the size of the printed parts, retrofitting the printed structure and existing porosity is of paramount importance that hinders the process. The material challenge is another obstacle in the field of building as there are no criteria to optimize the mixture concerning shrinkage, extrude-ability, and flow-ability to obtain the optimum mechanical properties [1].

Simulation is a valuable part of AM. It provides users with mathematical-based results. In turn, before any operations, users can decide whether their part or even process is accurate enough and investigate different designs before producing them. Engineers can also study the behavior of their design under different conditions and environments through simulation rather than expensive experimental tests. There are, however, some set of issues. As simulations are intertwined with high computing power, costly digital equipment and software licenses are needed. Due to the complexity of the design and components, simulation time might take longer than what is reasonable, and advanced computer systems are required to solve them [41]. The simplifying assumption would be an excellent choice to tackle this barrier. Being a relatively new technology, designers and engineers are not fully experienced, and as a result, simulation issues are of paramount importance in this regard, generating a false result [3].

On the other hand, 3D models are needed to analyze parts and generate the necessary codes to print out the parts. CAD software is the primary tool to do so, which encapsulates solid geometries and boundaries. Generating codes and transferring 3D models to form CAD software into a printed object can lead to inaccuracies and defects, especially when it comes to curved surfaces; therefore, post-processing might need to eliminate these effects. Choosing the optimum printing orientation and generating supporting pillars can be removed to address this issue. Moreover, assigning optimum printing parameters affects the mechanical properties of the printed parts, but it can also have a great influence on their appearance. Choosing an optimized parameter, then, is a challenge in this regard. In the case of the application when a flat surface is more desired, this challenge stands out [15]. Merging simulation and Big Data in Industry 4.0 can truly address this issue to select the optimized performance [41].

Combining Additive Manufacturing with other cyber aspects of Industry 4.0 can also manifest the possibility of the digital thread. All these digital threads can begin from the designing down to the customer phase, affecting optimized productivity and cost. Data generation, processing, and transferring are the basics of Industry 4.0. It means that there is a plethora of information processing and sharing through different chandelles. Using Big Data, Artificial Intelligence (AI), and Machine Learning can help the AM function properly. All these cyber aspects of Industry 4.0 need electronic devices to generate, transmit, and store data. As long as AM needs this information to execute tasks, the digital thread becomes much more critical. AM, in effect, needs a 3D model generated with CAD software and simulation to assess the product's behavior under different situations that can result in the development of a digital twin. Transferring data wirelessly through the Internet to the sensors, a digital twin can monitor the process and enhance the quality. The data transition via the Internet needs to be protected from any prospective cyber-attacks that affect the producing parts negatively and AM systems. Data protection from hackers to avoid losing confidential information is essential as well. Augmented Reality (AR) can also assess the physical parts and compare them with the 3D model during the process to identify any possible defects from the digital thread [41]. All in all, with all these flaws, Additive Manufacturing plays a crucial role in Industry 4.0.

7 Summary and Future Expectations

All three previous industrial revolutions have somehow changed the industries and manufacturing sectors. However, the last one, known as Industry 4.0, has brought integrated manufacturing systems using complex virtual information. AM, the layer-upon-layer manufacturing process, has brought a new era of high-quality mass and customized intelligent production. It helps time and cost-saving, reduced complexity, and improves the properties of the final product. The sustainability of AM is another development, which has been and will be achieved through less energy consumption and negative impacts on the environment. IoT, cloud computing, big data, etc., as Industry 4.0 components, have helped AM to be more efficient and intelligent. There are already some AM materials and process developments, and this trend seems to continue in the future. Industrial sustainability has been a preference for decades; companies and businesses have commenced investigating even more seriously how manufacturing can be performed in a more productive and environmentally sustainable way. Sustainable development guarantees that manufacturing operations result in less environmental, ethical, and economic impact.

In the future, decentralization might be possible by distributing the workload among factories/machines with the help of cloud services. Another future expectation is the sustainable development of AM, by which AM will play a vital role in lessening waste resources and decreasing energy consumption by using in-time production. On the other hand, society will benefit from 3D printing and smart manufacturing by redefining the role of employees and customers. Some challenges and barriers of on-location production will be solved with customized fabrication. Future smart materials along with innovative AM processes will be introduced and lead to high-quality products. AM has turned out to be an enabling technology, which reduces product design and development timelines effectively. That is a faster and cheaper manufacturing process. By implementing Industry 4.0 components, AM will become more intelligent in the near future. Rapid prototyping, printed bulky structures, reduced printing errors, and enhanced mechanical properties of products are the main factors of AM development, a technology that is still in its early stages.

References

1. Haleem A, Javaid M (2019) Additive manufacturing applications in industry 4.0: a review. J Ind Integr Manag 04:1930001. https://doi.org/10.1142/s2424862219300011
2. Dilberoglu UM, Gharehpapagh B, Yaman U, Dolen M (2017) The role of additive manufacturing in the era of Industry 4.0. Proc Manuf 11:545–54. https://doi.org/10.1016/j.promfg.2017.07.148
3. Dubey R, Gunasekaran A, Childe SJ, Bryde DJ, Giannakis M, Foropon C et al (2020) Big data analytics and artificial intelligence pathway to operational performance under the effects of entrepreneurial orientation and environmental dynamism: A study of manufacturing organisations. Int J Prod Econ 226:107599. https://doi.org/10.1016/j.ijpe.2019.107599
4. Mehrpouya M, Dehghanghadikolaei A, Fotovvati B, Vosooghnia A, Emamian SS, Gisario A (2019) The potential of additive manufacturing in the smart factory industrial 4.0: a review. Appl Sci 9. https://doi.org/10.3390/app9183865
5. Brant A, Sundaram MM (2015) A novel system for cloud-based micro additive manufacturing of metal structures. J Manuf Process 20:478–484. https://doi.org/10.1016/J.JMAPRO.2015.06.020
6. Kerbrat O, Le Bourhis F, Mognol P, Hascoët JY (2016) Environmental impact assessment studies in additive manufacturing. Environ Footprints Eco-Design Prod Process :31–63. https://doi.org/10.1007/978-981-10-0606-7_2
7. Diegel O, Nordin A, Motte D (2019) Additive manufacturing technologies, pp 19–39. https://doi.org/10.1007/978-981-13-8281-9_2
8. Aberoumand M, Rahmatabadi D, Aminzadeh A, Moradi M (2021) 4D Printing by Fused Deposition Modeling (FDM). Fused Depos. Model. Based 3D Print., Springer, Cham, pp. 377–402. https://doi.org/10.1007/978-3-030-68024-4_20
9. Firth J, Gaisford S, Basit AW (2018) A new dimension: 4D printing opportunities in pharmaceutics. AAPS Adv. Pharm. Sci. Ser., vol. 31. Springer Verlag, pp 153–62. https://doi.org/10.1007/978-3-319-90755-0_8
10. Khoo ZX, Liu Y, An J, Chua CK, Shen YF, Kuo CN (2018) A review of selective laser melted NiTi shape memory alloy. Materials (Basel) 11:11–17. https://doi.org/10.3390/ma11040519
11. Pyo Y, Kang M, Jang J young, Park Y, Son YH, Choi MC, et al (2018) Design of a shape memory composite(SMC) using 4D printing technology. Sensors Actuators, A Phys 283:187–95. https://doi.org/10.1016/j.sna.2018.08.049
12. Fonseca MA, Abreu B, Gonçalves FAMM, Ferreira AGM, Moreira RAS, Oliveira MSA (2013) Shape memory polyurethanes reinforced with carbon nanotubes. Compos Struct 99:105–111. https://doi.org/10.1016/j.compstruct.2012.11.029
13. Karlsson D, Lindwall G, Lundbäck A, Amnebrink M, Boström M, Riekehr L et al (2019) Binder jetting of the AlCoCrFeNi alloy. Addit Manuf 27:72–79. https://doi.org/10.1016/j.addma.2019.02.010
14. Zhou YH, Zhang ZH, Wang YP, Liu G, Zhou SY, Li YL et al (2019) Selective laser melting of typical metallic materials: An effective process prediction model developed by energy absorption and consumption analysis. Addit Manuf 25:204–217. https://doi.org/10.1016/j.addma.2018.10.046
15. Ngo TD, Kashani A, Imbalzano G, Nguyen KTQ, Hui D (2018) Additive manufacturing (3D printing): A review of materials, methods, applications and challenges. Compos Part B Eng 143:172–196. https://doi.org/10.1016/j.compositesb.2018.02.012
16. Melchels FPW, Domingos MAN, Klein TJ, Malda J, Bartolo PJ, Hutmacher DW (2012) Additive manufacturing of tissues and organs. Prog Polym Sci 37:1079–1104. https://doi.org/10.1016/j.progpolymsci.2011.11.007
17. Abedi HR, Hanzaki AZ, Azami M, Kahnooji M, Rahmatabadi D (2019) The high temperature flow behavior of additively manufactured Inconel 625 superalloy. Mater Res Express 6. https://doi.org/10.1088/2053-1591/ab44f6.

18. Chu C, Lin H, Liu H, Wang X, Wang J, Zhang P, et al (2017) Tumor microenvironment-triggered supramolecular system as an in situ nanotheranostic generator for cancer phototherapy. Adv Mater 29. https://doi.org/10.1002/adma.201605928
19. Sullivan TN, Pissarenko A, Herrera SA, Kisailus D, Lubarda VA, Meyers MA (2016) A lightweight, biological structure with tailored stiffness: the feather vane. Acta Biomater 41:27–39. https://doi.org/10.1016/j.actbio.2016.05.022
20. Miyashita S, Guitron S, Ludersdorfer M, Sung CR, Rus D (2015) An untethered miniature origami robot that self-folds, walks, swims, and degrades. Proc. IEEE Int. Conf. Robot. Autom., vol. 2015- June, Institute of Electrical and Electronics Engineers Inc., pp 1490–6. https://doi.org/10.1109/ICRA.2015.7139386
21. Gu GX, Wettermark S, Buehler MJ (2017) Algorithm-driven design of fracture resistant composite materials realized through additive manufacturing. Addit Manuf 17:47–54. https://doi.org/10.1016/j.addma.2017.07.002
22. Gu GX, Chen CT, Buehler MJ (2018) De novo composite design based on machine learning algorithm. Extrem Mech Lett 18:19–28. https://doi.org/10.1016/j.eml.2017.10.001
23. Zhi L, Hu YS, El Hamaoui B, Wang X, Lieberwirth I, Kolb U et al (2008) Precursor-controlled formation of novel carbon/metal and carbon/metal oxide nanocomposites. Adv Mater 20:1727–1731. https://doi.org/10.1002/adma.200702654
24. Holmes B, Zhu W, Li J, Lee JD, Zhang LG (2015) Development of novel three-dimensional printed scaffolds for osteochondral regeneration. Tissue Eng Part A 21:403–415. https://doi.org/10.1089/ten.tea.2014.0138
25. Zhu W, Holmes B, Glazer RI, Zhang LG (2016) 3D printed nanocomposite matrix for the study of breast cancer bone metastasis. Nanomed Nanotechnol Biol Med 12:69–79. https://doi.org/10.1016/j.nano.2015.09.010
26. Wang X, Jiang M, Zhou Z, Gou J, Hui D (2017) 3D printing of polymer matrix composites: A review and prospective. Compos Part B Eng 110:442–458. https://doi.org/10.1016/j.compositesb.2016.11.034
27. Rahmatabadi D, Aminzadeh A, Aberoumand M, Moradi M (2021) Mechanical characterization of fused deposition modeling (FDM) 3D printed parts. Fused Depos. Model. Based 3D Print, Springer, Cham, pp 131–50. https://doi.org/10.1007/978-3-030-68024-4_7
28. Aminzadeh A, Aberoumand M, Rahmatabadi D, Moradi M (2021) Metaheuristic approaches for modeling and optimization of FDM process :483–504. https://doi.org/10.1007/978-3-030-68024-4_25
29. Moradi M, Aminzadeh A, Rahmatabadi D, Hakimi A (2021) Experimental investigation on mechanical characterization of 3D printed PLA produced by fused deposition modeling (FDM). Mater Res Express. https://doi.org/10.1088/2053-1591/abe8f3
30. Sood AK, Ohdar RK, Mahapatra SS (2010) Parametric appraisal of mechanical property of fused deposition modelling processed parts. Mater Des 31:287–295. https://doi.org/10.1016/j.matdes.2009.06.016
31. Chohan JS, Singh R, Singh Boparai K, Penna R, Fraternali F (2017). Dimensional accuracy analysis of coupled fused deposition modeling and vapour smoothing operations for biomedical applications. https://doi.org/10.1016/j.compositesb.2017.02.045
32. Padhi SK, Sahu RK, Mahapatra SS, Das HC, Sood AK, Patro B et al (2017) Optimization of fused deposition modeling process parameters using a fuzzy inference system coupled with Taguchi philosophy. Adv Manuf 5:231–242. https://doi.org/10.1007/s40436-017-0187-4
33. Parandoush P, Lin D (2017) A review on additive manufacturing of polymer-fiber composites. Compos Struct 182:36–53. https://doi.org/10.1016/j.compstruct.2017.08.088
34. Utela B, Storti D, Anderson R, Ganter M (2008) A review of process development steps for new material systems in three dimensional printing (3DP). J Manuf Process 10:96–104. https://doi.org/10.1016/j.jmapro.2009.03.002
35. Lee H, Lim CHJ, Low MJ, Tham N, Murukeshan VM, Kim YJ (2017) Lasers in additive manufacturing: a review. Int J Precis Eng Manuf Green Technol 4:307–322. https://doi.org/10.1007/s40684-017-0037-7

36. Dou R, Wang T, Guo Y, Derby B (2011) Ink-Jet printing of zirconia: coffee staining and line stability. J Am Ceram Soc 94:3787–3792. https://doi.org/10.1111/j.1551-2916.2011.04697.x
37. Travitzky N, Bonet A, Dermeik B, Fey T, Filbert-Demut I, Schlier L, et al (2014) Additive manufacturing of ceramic-based materials. Adv. Eng. Mater., vol 16. Wiley-VCH Verlag, pp 729–54. https://doi.org/10.1002/adem.201400097
38. Gibson I, Rosen D, Stucker B, Gibson I, Rosen D, Stucker B (2015) Directed energy deposition processes. Addit. Manuf. Technol., Springer, New York, pp. 245–68. https://doi.org/10.1007/978-1-4939-2113-3_10
39. Williams SW, Martina F, Addison AC, Ding J, Pardal G, Colegrove P (2016) Wire + Arc additive manufacturing. Mater Sci Technol 32:641–647. https://doi.org/10.1179/1743284715Y.0000000073
40. Kazemian A, Yuan X, Cochran E, Khoshnevis B (2017) Cementitious materials for construction-scale 3D printing: laboratory testing of fresh printing mixture. Constr Build Mater 145:639–647. https://doi.org/10.1016/j.conbuildmat.2017.04.015
41. Butt J (2020) Exploring the interrelationship between additive:1–33
42. Salama M, Elkaseer A, Saied M, Ali H, Scholz S (2019) Industrial internet of things solution for real-time monitoring of the additive manufacturing process. Adv Intell Syst Comput 852:355. https://doi.org/10.1007/978-3-319-99981-4_33
43. Majeed A, Zhang Y, Ren S, Lv J, Peng T, Waqar S et al (2021) A big data-driven framework for sustainable and smart additive manufacturing. Robot Comput Integr Manuf 67:102026. https://doi.org/10.1016/j.rcim.2020.102026
44. Gutowski TG, Branham MS, Dahmus JB, Jones AJ, Thiriez A, Sekulic DP (2009) Thermodynamic analysis of resources used in manufacturing processes. Environ Sci Technol 43:1584–1590. https://doi.org/10.1021/ES8016655/SUPPL_FILE/ES8016655_SI_001.PDF
45. Singh M (Mrityunjay), Ohji T (Tatsuki), Asthana R (Rajiv) (n.d.) Green and sustainable manufacturing of advanced material
46. Gholami H, Abu F, Lee JKY, Karganroudi SS, Sharif S (2021) Sustainable manufacturing 4.0-pathways and practices. Sustainability 13:13956. https://doi.org/10.3390/SU132413956
47. Ford S, Despeisse M (2016) Additive manufacturing and sustainability: an exploratory study of the advantages and challenges. J Clean Prod 137:1573–1587. https://doi.org/10.1016/J.JCLEPRO.2016.04.150
48. Fraţila D, Rotaru H (2017) Additive manufacturing-a sustainable manufacturing route. MATEC Web Conf 94. https://doi.org/10.1051/matecconf/20179403004
49. Peng T, Kellens K, Tang R, Chen C, Chen G (2018) Sustainability of additive manufacturing: an overview on its energy demand and environmental impact. Addit Manuf 21:694–704. https://doi.org/10.1016/J.ADDMA.2018.04.022

The Impact of the Fourth Industrial Revolution on the Transitory Stage of the Automotive Industry

Mina Ahmadi, Mostafa Pahlavani, Armin Karimi, Mahmoud Moradi, and Jonathan Lawrence

1 Introduction

Industrial developments have been around for quite some time; however, as stated by Rifkin, whenever transportation and communication systems are redesigned, and the energy consumption is enhanced, the new revolution in the industry would begin [1]. More than 20,000 parts and components in a single vehicle, sourced from different suppliers worldwide, make the automotive industry's role in the economy incompatible with any other manufacturing industry [1, 2]. The importance of this sector is more evident by considering not just production and assembly processes but also sales and after-sales services, marketing, and maintenance of vehicles [3]. By being one of the most supplier-dependent and purchasing-dependent sectors, the automotive industry is one of the most complex manufacturers, which has gone under ever-present changes since its invention in 1886 [2].

Shortly after the first industrial revolution in the late 18th, which brought steam/water-powered processes, inventors started to test automobiles with steam-powered engines. Those engines could power vehicles to push pistons, turning the crankshaft and wheels through produced steam by heating water in boilers. However, so much added weight to a vehicle by employing steam engines—that were already

M. Ahmadi · M. Pahlavani
School of Automotive Engineering, Iran University of Science and Technology, Tehran, Iran

A. Karimi
Department of Aerospace Engineering, Sharif University of Technology, Tehran, Iran

M. Moradi (✉)
Faculty of Arts, Science and Technology,
University of Northampton, Northampton NN1 5PH, UK
e-mail: Mahmoud.Moradi@northampton.ac.uk

J. Lawrence
School of Computing, Engineering & Physical Sciences, University of the West of Scotland (UWS), High Street, Paisley PA1 2BE, UK

H. Gholami et al. (eds.), *Sustainable Manufacturing in Industry 4.0*,
https://doi.org/10.1007/978-981-19-7218-8_5

dealing with poorly structured roads—led to the invention of internal combustion engines and the Otto four-stroke cycle. That invention was then applied to gas and petrol engines after the second industrial revolution. After the first industrial revolution and almost a century later, the transition from the first to the second industrial revolution with extensive advances in the industry that contributed to the emergence of new energy sources (electricity, gas, and oil) took place. Other highlights of the second industrial revolution were the growing demand for steel, chemical synthesis, and communication methods such as the telegraph, wireless radio, and telephone. Mass production based on electric power and automotive conveyors is among the significant accomplishments of this industrial revolution. It took another century to witness the third industrial revolution in the second half of the twentieth century with the advent of another energy source called nuclear energy that was previously unusable and microprocessors. The third industrial revolution saw the beginning of electronics, telecommunications, and, of course, computers. The third industrial revolution opened the door to space exploration, research, computerized systems, and controls in manufacturing processes by utilizing new technologies. Also, in the industrial world, two great inventions of this period, programmable logic controllers PLCs and robots, have helped usher in the age of high-level automation [1, 2].

Today, the term Industry 4.0, also known as the fourth industrial revolution, which was first used by a German professor, Klaus Schwab, in 2011, defines cyber-physical systems that enable the vision of intelligent machines controlling themselves during manufacturing processes through analyzed collected data. This information is gathered and interpreted via the Internet, big data analytics, cloud computing, etc., and other types of software and hardware which have been created during the third industrial revolution and have been developing significantly [4].

This revolution is the age of intelligent devices, storage, and production systems that can independently exchange information, perform operations, and control equipment without human intervention. In the corporate world of the twenty-first century, Industry 4.0 stands for a new era in manufacturing processes based on the interconnection of data flows among different actors (e.g., manufacturers, suppliers, and consumers) and vertical/horizontal intelligent linking of production processes inside organizations from the first stages to finished product [5].

Despite the many advantages that Industry 4.0 brings, it also causes some social and environmental concerns that make sustainable development more crucial than before. Industry 4.0 must fit into ecological and social issues without disregarding the importance of economic growth. As one of the largest industries in the world that is a pioneer in adopting Industry 4.0, the question is, "what is the role of sustainable development in integrating the concept of Industry 4.0 into the automotive sector?" [6]. Sustainability design balances ethical, environmental, and social issues and economic factors within the product and service advancement process. That ensures that all business and society needs are met while the ecosystem is protected. The principles of the Design-for-X methodology, established by Jawahir et al., can be exploited to produce a sustainable product. These can help investigate and analyze the vehicle and its sub-systems design procedures and their impacts on the economy, society, and environment. Designs for manufacturing, recyclability, minimizing material usage, durability, maximizing energy efficiency, and minimizing greenhouse emissions are subsections of DfX rules [7].

The automotive industry is in the middle of a green transformative revolution driven by innovations in cloud technology. In a world where users demand a seamless purchasing experience, passion for mobility options, and personalized cars as their smartphones, we can anticipate the speed of smart technology and the Industry 4.0 sub-systems adoption in the automotive sector to stimulate. The durable trends of sustainability, security/data protection, convenience, and personalization are consistent, but technology is driving and changing these trends every day.

In this chapter, the evolution of cyber-physical systems enabling digitalization in the automotive industry under the umbrella of Industry 4.0 is thoroughly discussed. Sustainable development and the challenges in the way of digitalization in the automotive sector, as well as the future of the smart automotive industry, are also put in the following sections.

2 Evolution of Embedded Systems to Cyber-Physical Systems in the Automotive Industry Through Industry 4.0

Due to its flexibility and versatility, an embedded system (ES), a complex set of microprocessor/microcontroller-based hardware and software systems, has been getting control of many industrial functions. This system is exclusively designed for performing dedicated functions either as a part of a whole mechanical/electronic or independent system. The basic structure of such a system includes sensors, actuators, analog-to-digital, and digital-to-analog converters, which were introduced during the third industrial revolution [8].

ES has presented a new configuration of an automatic closed-loop and real-time controller in the automotive industry since its introduction in the 1960s. By turning to be the seamless integral part of vehicles, ESs have mainly enhanced automobiles' performance, efficiency, and functionality after the advent of semiconductor technology categorized as sensors, memory devices, microcontroller units, and transceivers employed at different levels of applications. Airbags, anti-lock braking systems (ABS), emission control, tire pressure monitor, climate control, navigational systems, satellite radio, etc., are all examples of embedded systems in automobiles to fulfill the demands for safety and comfort, optimized fuel consumption, and pollution reduction. Even though ESs have accelerated the technological transition of the automotive industry in recent years, the rapid continuous development of technology alongside modernization, artificial intelligence, vehicle electrification, and increased awareness toward global warming have driven these systems to evolve into cyber-physical systems (CPS) in Industry 4.0. The embedded cognitive system will be at the heart of such transition, and CPSs communication and internal/external direct control of physical or digital processes are enabled by Industry 4.0 using information technology [8, 9].

Although Industry 4.0 is based on embedded and cyber-physical systems by which the collected data come to help machines and production sectors control themselves, it also stands for a new level of mechanical intelligent manufacturing processes, including Additive Manufacturing (AM) and Robots.

The roles of each Industry 4.0s key components in the automotive industry's digitalization are explained in the following sections.

2.1 *Robots in the Automotive Manufacturing*

Today's world is shifting to automation, and robots, machines with the capability to handle a complex series of tasks automatically, have found enough significant applications in intelligent manufacturing to impact the value chain, especially in the automotive industry as a pioneer in the industrial usage of robots. These applications vary from welding to painting, assembling, sealing, removing materials, coating, and transferring parts. Pick-and-place robots, for instance, lead to an increased production rate by picking up and replacing components at speeded rates. Or robots with long arms and higher payload capabilities handle spot welding on heavy body panels. As an industrial sector with the largest number of applied robots, the automotive industry utilizes robots to increase production speed, achieve higher levels of accuracy, reduce labor costs, and protect employees from difficult and dangerous tasks [1, 4].

A Tier[1] 1 supplier, Brose Ostrava employs conveyor/pick-and-place robots in material-handling applications to transport the daily received material from more than 200 suppliers and machines loading and unloading in combination with automated warehouses. Handing such repetitive, hard tasks over robots lets operators and engineers focus on other tasks with higher importance and result in higher productivity, similar to what is happening in Continental Tier 1 supplier, which employs robots to operate alongside humans for carrying PCB boards. Today's autonomous robots in the automotive sector could be fed by engineers, operators, and cloud-based systems and be controlled remotely [1].

2.2 *The Value of Additive Manufacturing/3D Printing in the Automotive Industry*

Constant competitive challenges among Original Equipment Manufacturers (OEMs) and their suppliers to achieve higher performance standards have driven them to

[1] Companies that supply parts or systems directly to Original Equipment Manufacturers.

employ high-tech manufacturing processes to overcome the problem of fuel efficiency, time consumption, cost, aerodynamics (in vehicles), innovation, safety, security, and connectivity. To fulfill one of the basic demands of customers, customization and fabrication of complex freely designed objects with advanced and most influential attributes, additive manufacturing (AM) became one of the important parts of Industry 4.0. Significant advances in Additive Manufacturing/3D printing and its subsets, rapid prototyping,[2] and rapid tooling[3] have turned this technology into a promising method to transform the potential ways of designing, testing, and manufacturing [4, 10]. AM is building up a 3D-designed part layer by layer through a controlled material deposition. However, this technology has paved the paths for mass production of high-performance products in the automotive industry, designed scale models for testing procedures, passed rigorous verification, and finally, lighter and safer products, shorter lead time, and lower costs achieved [11]. Original Equipment Manufacturers and suppliers in the automotive sector benefit from rapid developments and innovations in AM and advanced materials used in 3D printers. The automotive industry is now adopting the new strategy of AM to go through the challenge of freedom of design for the advent and production of complex but lightweight components.

In the design phase of the production cycle, companies experience some difficulties before deciding on the final design. But one of the most outstanding merits of AM, enabling producing multiple variations of products with fewer additional costs and design restrictions, helps automakers enhance their products' design by benefiting from physical models and prototypes at higher speeds. Clay models are progressively replaced by CAD-designed files, which are then converted to 3D prototypes.

Failure or malfunction of a product or part may result in catastrophic consequences. Therefore, a new component must be thoroughly evaluated to determine whether it is performing as intended or not before sending it to the market. For these purposes, rapid prototyping is an ideal option in the modern world of industry. By employing AM for rapid prototyping, testing the quality of the desired outcome will be available before the final stage of production through building prototypes. By gaining the advantages of minimizing inventory and avoiding overproduction, more experiments and prototyping can be performed by AM to meet the customer's and suppliers' needs, resulting in better customized and authentic products.

In the automotive industry, tooling's[4] role on the assembly line is prominent to obtain high-quality and customized tools. However, for some automobiles' components, tooling and investment casting are time-consuming and too expensive processes. By using AM in the design phase, automakers can reduce their dependence on tooling and casting. Additionally, improving fuel efficiency can be achieved via AM's ability to use light materials to produce lightweight structures without missing

[2] Fast fabrication of a physical part, model, or assembly using 3D computer aided design (CAD).

[3] When Rapid Prototyping techniques and conventional tooling practices are used together to produce a mold quickly.

[4] Building the different types of components and machinery needed for production, like molds, jigs, and fixtures.

the strength. By exploiting the AM, automakers are now using additive technologies like Selective Laser Sintering (SLS) and Selective Laser Melting (SLM) to print different end-use components, including custom spoilers, windbreakers, bumpers, and other cars' parts. SLM is also employed to fabricate emission systems from heat-resistant Al alloys. In addition, printing pumps and valves are now available via sophisticated AM technology of Electron Beam Melting (EBM) [10].

2.3 The Internet of Things (IoT) Trends in the Automotive Industry

When it first emerged, the Internet enabled the connection between people and organizations worldwide. Still, there is now a talk on the Internet of Things, shaping Information and Communication Technologies development (ICTs) using the networked interconnection of everything in everyday life [1, 3]. As an emerging paradigm, IoT makes people's day-to-day life easier and more convenient by providing platforms for devices to communicate their physical context information like location or status to the Internet, other objects, machinery, or even humans [12]. Moreover, beyond the existing barriers between the physical and digital worlds, IoT lets physical objects identify their surroundings, interact with humans, and even make decisions and become self-controlled [13]. Since some technologies such as RFID,[5] NFC,[6] or Sensor and Actuator Networks and even mobile phones cannot be considered as a novelty, rather than being a revolution, IoT is an evolution. However, by being embedded in Industry 4.0, IoT evolves more, experiencing an infrastructural revival in the types and number of devices it employs and how they are connected [1].

The automobile industry is on the brink of a revolution, and as one of the most critical industries, a remarkable transformation that is about to happen in this sector from human-guided to automated guided/self-driven and connected vehicles will have a long-term impact on our daily lives [8]. Of course, automobiles have been linked to smartphones, registered real-time traffic alerts, employed GPS and navigation systems, etc., for quite some time. Still, the vision of autonomous and connected cars is becoming a reality with the Internet of Things. This transformation will shift the automobile industry to the age of services and experiences from products, to software from hardware to information from functionality, and complex and connected ecosystems from industry silos [8, 9].

Connected cars can interact with their surrounding environment, which means they can communicate and transfer data to other vehicles and external devices and infrastructures using sensors and either local wireless networks or the Internet. Extrapolating the IoT concept to the automotive industry, it can be summarized into three

[5] Radio Frequency Identification (RFID) refers to a wireless system comprised of two components: tags and readers.

[6] Near-Field Communication (NFC) is a set of communication protocols for communication between two electronic devices over a distance of 4 cm (11/2 in) or less.

connection branches. The first one is the connectivity of one vehicle to one or even more vehicles (Vehicle-to-vehicle (V2V)); the ability to wirelessly exchange information about the speed and position of surrounding vehicles shows great promise in helping to avoid crashes, ease traffic congestion, and improve the environment. The second is the connection between cars and external infrastructures (vehicle-to-infrastructure (V2I or v2i)). This communication model allows vehicles to share information with the components that support a country's highway system, such as cameras, traffic lights, parking meters, and street lights. And finally, the third is a link between vehicles and external hardware or devices (vehicle-to-devices (V2D)), a particular type of vehicular communication system that consists of the exchange of information between a vehicle and any electronic device that may be connected to the vehicle itself [14].

The complete picture of connected cars and, as a result, autonomous vehicles can be achieved by managing the collected and analyzed data and connecting everything to the Internet. Safety and security systems of cars, telematics, and in-vehicle-infotainment are improving by benefiting from IoT [15]. Automatically transferring real-time data through the Internet, connected cars can send their locations to emergency teams in case of accidents (emergency call). Connected smart cars can employ IoT, integrate GPS with online services, and utilize them for driver preferences, navigating, routing, fuel station availability, traffic alerts, the expiration date of insurance alarm, avoiding a traffic jam, etc. Road conditions, other surrounding vehicles information, and in-time-diagnosis of car problems are examples of IoT beneficial aspects in improving the safety of vehicles and roads. Remote control parking, stolen vehicle recovery, online in-vehicle entertainment options, etc., are also the results of embedding IoT into the automotive industry [15].

By sensing and detecting physical objects around a vehicle using built-in smart sensors and processing the received external data via algorithm software, the driving software will define the speed and direction of the car. This is what is expected to happen in autonomous cars based on the technology of connected cars using IoT.

Figure 1 shows the stages of connected cars' evolution from the mid-1960s to the new mobility era going beyond 2020 based on Industry 4.0.

Phase one was the Research and Development (R&D) era, the longest; during nearly 30 years, great innovative ideas were proposed, but the lack of technology didn't let them be implemented. Next was the era of an embedded digital communication module (DCM), e.g., mobile phones and sensors used for transferring and communicating data and information wirelessly to an automaker or TSP (telematics service provider). The infotainment (information and entertainment) era started in 2007 and lasted for five years, introducing applications based on information sharing and entertainment within a car. A shift of power equation in industries, third-party apps, software providers, and app providers are engaged in this stage.

Vehicle-to-vehicle or vehicle-to-infrastructure communication became possible by integrating electronic gadgets, smart devices, and sensors in the era of V2X integration started in 2012. And finally, the new mobility era, the stage of autonomous cars' emergence and evolution, the age of embedded Industry 4.0 components in the automotive sector [8]. As an Original Equipment Manufacturer, Volkswagen uses

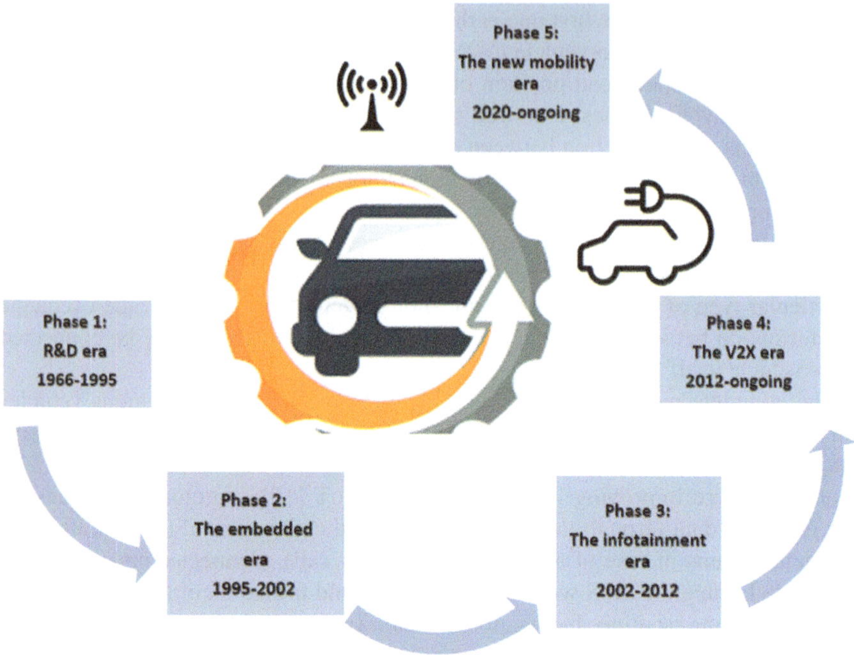

Fig. 1. 5 stages of connected vehicles' evolution path

IoT for monitoring returnable packages and load carriers across the supply network. As a result, suppliers can track components from the first point to their destinations through end-to-end visibility [1]. To be more specific, Smart manufacturing refers to data centers that are available to the users through the Internet conveniently [16]. The Automotive sector, with its complex requirements, is completely distinguished from other manufacturers. However, its ongoing revolution around designing, engineering, simulations, data, analytics, and dealers' network require compelling on-site IT infrastructure whose maintenance seems to be a technical challenge. With the help of its super servers, facility of transferring and processing real-time data, cloud computing can help this industry overcome the issue of the flow of information and data. The best example of the automotive industry taking advantage of technology is Google's self-driving car, which changed the impossible into the possible thing [17]. In case of enhancing and leveraging the safety and security of vehicles, cloud computing helps IoT predict mechanical failures and nullifies the downtimes, alerts drivers about disruption in the road condition, etc., based on shared information and resources via V2D, V2I, and even V2V communications [18]. The automotive industry can benefit from cloud computing due to its faster, better, and safer data processing and storage. As the leading automotive Tier 1 supplier, Bosch set up its Cloud Backup from Team Knowhow as an easy way of keeping its files and information backed up securely online. Another example is a private cloud of Volkswagen, which provides dedicated applications for its internal operational sectors, clients, suppliers, and sales organizations [1].

2.4 *Impact of Big Data and Analytics on Automotive Industry*

Big Data is characterized as high-variety, large-volume, high veracity, and high-velocity datasets, and analyzing them using traditional processing procedures is hardly attainable [1, 19].

While volume refers to the quantity of huge and growing data on a daily basis, velocity represents the speed of generating and transferring the data. The term variety refers both to the different sources from which data is gathered and various data types. Finally, veracity means checking the reliability and accuracy of a large amount of data that comes from different sources. 4vs of big data are shown in Fig. 2.

Internet is overgrowing these days, which means a larger amount of generated data should be gathered daily. Big data technology can analyze and separate data based on their criticality, which means more critical data from less critical ones. A clear picture of the situation based on interpreted data is transferred by this technology to fulfill business activities.

Big data analytics has become mandatory for all industries and domains because of the rapid development in networking, and a higher amount of collected and stored data through IoT and cloud computing. However, the emphasis is on what an organization does with such data rather than the data volume. As an essential analytical basis for insights, big data leads to better decisions and strategic directions in business [3].

Today's high-tech era needs a 24 × 7 connection between automakers and users, markets, and corporations through access channels like IoT and cloud computing.

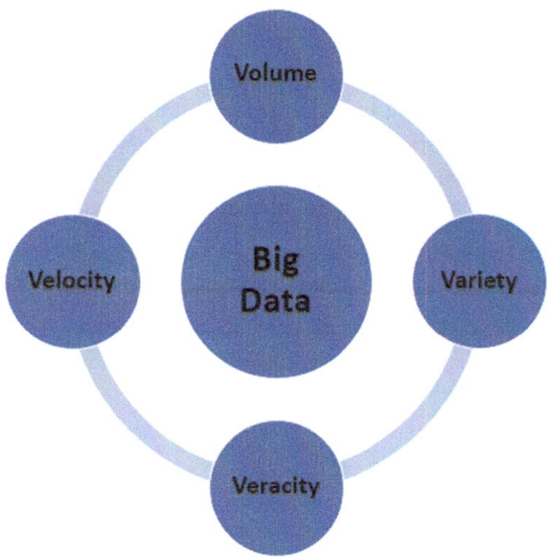

Fig. 2. 4vs of big data

The received data is enormous, and automakers are keen to employ new dealing methods to collect, analyze, and assist them in acting on such data. Getting a deep insight into the continuously obtained data from external resources via IoT is sometimes banned due to data silos. A certain automaker needs to fully understand their customers' priorities and place them on the agenda and adopt better strategies to configure the optimum incentives and marketing for the targeted group of customers [20]. Data management/Big data analytics is highly recommended for automotive industry players to protect and keep their connection with their customers. To target their customers ' priorities, they need to keep an eye on their sales, marketing, information services, and the gathered data through apps, social media, V2V, V2D, and V2I connections. Exploiting data from modern connected cars to meet customers' needs and optimize their experience by automakers can be done by employing systems capable of capturing data from customers' experiences through different communication channels and getting a clear view of their preferences, analyzing the stored data, and finally making decisions on what is required to serve their goals effectively [20]. Such big data sources vary from social networks (Twitter, Facebook, blogs, comments and pictures, YouTube, Instagram, etc.), business and commercial transactions, and IoT (traffic control, weather forecast systems, security cameras, and computers Internet-based systems). When conventional traditional database tools become incompetent in handling a large volume of data with great variety and velocity, big data comes into play. Hidden markets are then exposed, customer preferences are found, and cost reduction is achieved by an appropriate big data analysis.

Some OEMs are now using big data analytics for optimizing transportation networks such as transportation times, routing, and truckloads; forecasting transportation delays; and setting alarms [1].

2.5 Blockchain and the Automotive Industry

Recent advances in connected and autonomous vehicles (CAVs) made them attract more attention; however, due to safety, privacy, and security obstacles and concerns, a great uncertainty is around the re-blooming industry of the automotive. The problem is that these vehicles' operation basically depends on the online network and is susceptible to different software and hardware faults and cyberattacks. Different networks model (interconnections among entities and users) such as decentralized[7]

[7] A decentralized network has numerous connections between nodes. However, there is still the possibility of connection lost within the network.

and distributed[8] networks have been proposed as a substitute for centralized[9] ones. Among them, blockchain has become one of the most promising models [1, 21].

A blockchain is defined as a tamper-free/open distributed ledger capable of storing transactions among different parties exchanged via networks in an efficient verifiable way. With these abilities, this technology can make the vehicle-related experience much better by enabling several applications. However, the automotive industry is among those industries that stand to benefit from the technology of blockchain. Blockchain can lend itself to expanding the functionality of Connected and automated vehicles (CAVs) by improving their security, enhancing users' privacy, and increasing the safety of passengers. Furthermore, the history of records on a vehicle's performance and its parts is vital for a vehicular industry; for example, this recorded data can provide information for insurance and compensation actions in accidents.

Another example is odometer fraud detection; blockchain can record mileage information online, making odometer tampering easily recognizable. Daimler (a German multinational automotive corporation) joined to use the Blockchain in Transport Alliance (BiTA) to create blockchain standards in freight transportation. With the help of blockchain, it will be easy to track the automotive components through the supply chain, improve maintenance records, and recall products more precisely [22].

BMW is another OEM that employs blockchain to ensure the cobalt used in its battery is clean and is not mined by artisanal miners or even children under the risk of health issues or human rights abuses [1].

2.6 *How Virtual and Augmented Reality Are Changing the Automotive Industry*

Virtual Reality (VR)[10] and Augmented Reality (AR)[11] have entered the world of engineering recently. Due to the pressure of reducing time-to-market, increasing productivity and reliability, and quality of products, the automotive industry has become the leading industrial sector in employing VR/AR in different processes, from designing to manufacturing, testing, and even training [23, 24]. By virtualization, manufacturers can use cyber-physical systems and receive information from sensors and other components to create virtual models that represent the physical world. Both VR and AR are widely used in logistics due to the improved efficiency

---

[8] A distributed network has several connection paths among nodes, and the possibility of disconnectivity has been reduced drastically.

[9] A centralized network has one central node, which is several connection paths that have been derived from it.

[10] A system allows users to feel they are in the real world by interacting, moving, and being immersed in a 3D environment.

[11] A system that uses virtual simulations for representing design workplaces and different processes.

they provide processes with. Logistics play a fundamental role in automotive companies by cooperating with different production processes, ensuring the supply chain of parts, and presenting the realization projects with the help of AR and VR.

Primary applications of AR/VR in the automotive sector are classified as follows [23, 24]:

- **Design**: Car designing is an expensive and time-consuming process that needs continuous reviews and modifications before reaching the final design. However, VR/AR can reduce the time and cost by substituting virtual mock-ups for physical ones. Moreover, they bring simplification in case of trial and error by avoiding rebuilding mechanical/physical parts.
- **Virtual Prototyping (VP)**: It is a subsection of the designing process and, with recent advances, can be used to replicate physical models with less time and cost. By developing VR and VP and different software and hardware, models could be modified easily during the first stages of the production.
- **Manufacturing**: Virtual Manufacturing (VM) is a term that describes using VR or computers for improving decision-making abilities, enhancing risk measures, and controlling the manufacturing processes more effectively.
- **Virtual Assembly (VA)**: It provides assembling and disassembling of the virtual components, for example, how and where to mount parts in a vehicle.

Another usage of VR is a virtual simulation of crash tests and other phenomena such as night driving. With the ability of VR in virtual representations of the conditions of workplaces, evaluation of employees' health, safety, and well-being measures is available now.

Jaguar Land Rover (JLR) is the world's leading automaker implementing VR technologies for various automotive applications, including Cave Automatic Virtual Environment (CAVE). That is a virtual reality space where the automobile's parts act as giant projection surfaces to create a highly immersive virtual environment [24].

2.7 *Artificial Intelligence and Deep Learning in the Automotive Industry*

Artificial intelligence (AI) is a broad field of computer science with multiple approaches. One of the most prominent applications of this technology is automating the driving process. Since automated driving has gained much interest in the last decades, much academic and industrial research has been conducted so far. This field is a relatively new area of industry with some advantages as well as disadvantages. However, the legislature of many countries put some limitations on this technology.

One of the essential demands of this technology is an advanced sensor. Nowadays, three sensors are being utilized in semi-autonomous cars: cameras, RADAR, and LIDARS. For instance, CMOS is a kind of optical sensor (camera) that is widely used to aid in state-of-the-art self-driving cars. Furthermore, many researchers and engineers are trying to design more enhanced sensors. One way to improve the sensors is sensor fusion; the combination of some sensors will help self-driving cars better. Intelligent control systems in Self-Driving cars are launched by artificial intelligence. This science focus on the actual route planning process. That is, how the vehicle will move about its environment and the dynamic obstacles that it encounters. The car must have the ability to lane detection, object detection, and object classification. For this aim, the Neural Networking model, machine learning, and machine vision are performing. Apart from the design of Self-Driving cars, AI plays an essential role in manufacturing process optimization in the automotive industry. Performing Collaborative Robots, Automated Guided Vehicles, Painting Robots, and Automated Welding are examples of using AI technology in the industry, which can help to increase efficiency [25, 26].

Deep learning, also known as the deep neural network, is one of the most prominent branches of artificial intelligence (AI) that mimics how humans obtain certain kinds of knowledge. It is proven that performing deep learning in the automotive industry is very beneficial. For instance, one area that recently saw a considerable improvement in deep learning is computer vision. Also, deep learning has several well-known applications in image analysis and image classification. Generating an appropriately large amount of dataset for training networks and new tools and infrastructures for computation are important for this method. For the aim of data storage and processing, cloud computing is increasingly becoming a viable platform. Moreover, deep learning benefits online services such as image classifying in Google. The fact that social data is highly unstructured, known as big data, makes deep learning a precious tool for businesses to manipulate data. These companies use deep learning to decide which concepts might be of interest to which customers [26].

3 Sustainable Development in the Automotive Industry

With today's globalization and digitalization, countries and, particularly, organizations and manufacturers face sustainability challenges. That means they need to be more socially and environmentally responsible and protective while seeking high economic performance, referring to the term "sustainable development." A straightforward definition of sustainable development is using/reusing/recycling resources to satisfy present needs without negatively impacting the next generations using these resources to meet their own needs. Sustainability guarantees long-term business success and improves living standards by balancing social, ethical, and environmental concerns and economic factors during the production processes [7]. The sustainable development of the automotive industry is defined as human and planet-friendly processes and operations, final products, and services.

Established by Jawahir et al., Design-for-X is a framework exploited to design a sustainable product by analyzing and considering the manufacturing process's impact on the environment. DfX principles can be applied to the following subsections in the automotive industry:

- Design for Manufacturing (DfM): This methodology constitutes a number of guidelines that are not limited to the product's different structures, cutting lead time and production cost, adopting the product at the company level, etc. One deviator of this subsection is the Design for Assembly (DfA), which aims to focus on fastening and assembling guidelines.
- Design for recyclability: It includes three designs for remanufacturing, recycling, and disassembling the parts. That is ensuring the minimum time and costs of disassembling the vehicles' parts (disassembling), returning those parts to an appropriate level of performance while reducing the waste (remanufacturing), and finally processing materials out of one form and changing them to another new product (recycling).
- Design for minimizing material usage: It is a set of strategies for decreasing the amount of material used over the production life cycle and reducing its adverse environmental impacts.
- Design for durability: That is ensuring that the product will not fail during a specific functioning period.
- Design for energy efficiency: It is reducing fuel consumption and greenhouse emissions by lightening the vehicle's weight, improving engine performance, and finding alternative renewable and green sources of energy [27].

Some major elements of sustainability in the automotive sector are as follows [27]:

- Sustainable R&D and engineering, which refer to designing products that aim to develop lightweight products with improved aerodynamics while using renewable and recyclable raw materials. Volkswagen is one of the industries employing raw materials, such as cotton and natural fibers for different components' production. BMW is another example of an automaker that employs sustainable design under economic policies. This company is substituting more plastic parts for metallic parts to decrease the weight and so increase fuel efficiency.
- Product's sustainability, that is moving to electric cars and even biodegradable parts. Electric and automated cars introduced by implementing CPSs of Industry 4.0 into the automotive sector contribute to sustainable development by lowering the released amount of greenhouse gases, decreasing air pollution, and providing new job vacancies, which is leading to positive social impacts.
- Sustainability in the supply chain is adopting eco-friendly operations in different logistics, warehousing, and distribution. For example, Scania is using bio-gas fuelled trucks and building gas filling stations.

4 Challenges of the Digital Transformation of the Automotive Industry

The automotive industry is about to meet a massive upheaval. Its transformation will be more extensive than ever before in its over 125-year history. Some critical challenges that automotive industries are facing in the way of intelligent transformation are as follows [1, 28]:

- Standardization—which its importance is often disregarded—is considered as a prerequisite for digital transformation in the automotive industry. However, due to the higher speed of technological developments that cause standards to fall behind and the lack of international/global standards in transportation systems, OEMs create their standards and force them on suppliers. Therefore, standardization challenges should be solved by paying more attention to funding standardization bodies and improving international cooperation relationships.
- For different companies to remain competitive in the world of Industry 4.0, they have to trust in and accept technologies such as IoT and blockchain in their companies. This acceptance and trust are nothing without data security. Therefore, ensuring data security is now a significant challenge and becomes more crucial in the future as the volume of the transferred data increases every second.
- Industry 4.0 and digitalization are considerably affecting the lives of employees. Working condition/environment is swiftly changing and becoming more digitalized soon which needs more trained specialists and operators which impose a cost and time burden on employers to train them. It also causes reductions in the number of staffs, which is not according to the social aspect of sustainability and probably will negatively affect society with an increased number of un-employments.
- While new technology adoption seems accessible, positive perspectives of employing high-tech devices are negatively affected by imposed huge costs of adopting and maintaining such intelligent machines and Internet-based infrastructures, especially for small- or medium-sized businesses.
- As one of the industries that are constantly struggling with environmental issues (e.g., CO_2 emission), the automotive industry is now facing the problem of enhancing its sustainable development. They also have to deal with social sustainability due to changing working environment by employing different smart devices and increasing unemployment rates because of robots.
- Finding renewable energy sources and using them as fuels in vehicles is another issue that should be addressed in terms of the automotive industry's green transformative.

5 Summary

Digital transformation actions should address a company's all business model as a cross-cutting issue and impact all key business processes. Therefore, a digitization policy that incorporates all areas should not be treated in isolation but should be developed as an indispensable element of a long-term, strategic corporate planning process. The revolution of intelligent devices, storage systems, and production equipment, known as Industry 4.0, continuously changes the world and industries. As one of the world's largest companies, the automotive industry is affected by this intelligent transmission from the third to the fourth industrial revolution. This chapter introduces the automotive industry's evolution through history, from the first industrial revolution to the latest one. Different intelligent elements of Industry 4.0, including IoT, big data analytics, cloud computing, blockchain, virtual reality, and artificial intelligence, and their influences on the automotive industry, are described in this chapter.

With ever-increasing fuel prices, scarcity of fossil fuel resources, and increased awareness toward climate change, consumers' and businesses' attention have been drawn to sustainable products. Nowadays, the automotive industry is considered the most prominent industry globally due to its environmental impacts and Co_2 emissions. Sustainability in the automotive industry and its concept, which is now an inseparable part of the manufacturing processes, is fully explained in this chapter.

It was concluded that although Industry 4.0 brings many advantages to the sustainable development of the automotive industry, it causes some challenges such as social sustainability by utilizing robots which are leading to redundancy in workplaces.

The summarization of the four pillars of digitalization in the automotive industry is as follows:

- **Connected vehicles and services**: These are available to consumers through infotainment services in the car or on their smart devices.
- **Mobility services**: These are autonomous cars and robots, which a human controls.
- **Sustainability and efficiency improvement**: That is, making companies and products more productive, sustainable, and efficient.
- **Customization**: That is establishing new strategies and marketing models according to customers' latest offerings.

Integrating the automotive industry and intelligent components of the fourth industrial revolution brings some challenges, such as cyber-security, which all are introduced in this chapter. Finally, the future of the automotive industry under the umbrella of Industry 4.0 is developed based on the concept of autonomous and intelligent vehicles. However, we are still far from the idea of smart cities and transportation systems due to the limitations and challenges of digitalization.

References

1. Kern J, Wolff P (2019) The digital transformation of the automotive supply chain -an empirical analysis with evidence from Germany and China: Case study contribution to the OECD TIP Digital and Open Innovation project
2. Nyamwange J, Nyamwange MK (2014) Major innovations in transportation: evolution of automobiles
3. Pezenatto L, Coti-Zelati P, Araújo D (2020) Industry 4.0 and sustainable development in the automotive sector indústria 4.0. 2020. 10:26–54
4. Kandasamy J et al (2020) Sustainable manufacturing for industry 4.0—an augmented approach
5. Gottge S, Menzel T, Forslund H (2020), Industry 4.0 technologies in the purchasing process. Ind Manag & Data Syst, ahead-of-print
6. Vaz CR, Rauen TRS, Lezana ÁGR (2017) Sustainability and innovation in the automotive sector: a structured content analysis. Sustainability 9(6):880
7. Mayyas A et al (2012) Design for sustainability in automotive industry: a comprehensive review. Renew Sustain Energy Rev 16(4):1845–1862
8. Krasniqi X, Hajrizi E (2016) Use of IoT technology to drive the automotive industry from connected to full autonomous vehicles. IFAC-PapersOnLine 49(29):269–274
9. Canuto da Silva G, Kaminski P (2016) From embedded systems (es) to cyber-physical systems (cps): an analysis of transitory stage of automotive manufacturing in the Industry 4.0 scenario
10. Javid, I. and S. Christmann, *Additive Manufacturing in the Automotive Industry*. 2020.
11. Chua K, Leong KF, Liu Z (2015) Rapid tooling in manufacturing, 2525–2549
12. Sinay J, Kotianová Z (2018) Automotive industry in the context of industry 4.0 strategy. In: Transactions of the VŠB: technical University of Ostrava, safety engineering series 13:61–65
13. Qin E et al (2013) Cloud computing and the internet of things: technology innovation in automobile service. in human interface and the management of information. Information and interaction for health, safety, mobility and complex environments. Berlin, Heidelberg: Springer Berlin Heidelberg
14. Arce-Plevnik L, Cohen A, Shor T (2016) IoT in automotive industry: connecting cars
15. Rajesh S et al (2018) Internet of things in automotive industries and road safety, in internet of things in automotive industries and road safety. River Publishers. p i–xxi
16. Whaiduzzaman M et al (2014) A survey on vehicular cloud computing. J Netw Comput Appl 40:325–344
17. Boukerche A, De Grande RE (2018) Vehicular cloud computing: architectures, applications, and mobility. Comput Netw 135:171–189
18. Ashokkumar K et al (2015) Cloud based intelligent transport system. Procedia Comput Sci 50:58–63
19. Gottge S, Menzel T, Forslund H (2020) Industry 4.0 technologies in the purchasing process. Ind Manag Data Syst 120:730–748
20. Singh S, Jagdev G (2020) Execution of Big Data analytics in automotive industry using hortonworks sandbox. In: 2020 Indo—Taiwan 2nd international conference on computing, analytics and networks (Indo-Taiwan ICAN)
21. Abbade LR et al (2020) Blockchain applied to vehicular odometers. IEEE Netw 34(1):62–68
22. Demir M, Turetken O, Ferworn A (2019) Blockchain based transparent vehicle insurance management. In: 2019 sixth international conference on software defined systems (SDS)
23. Čujan Z, Fedorko G, Mikušová N (2020) Application of virtual and augmented reality in automotive. Open Eng 10(1):113–119
24. Lawson, G., D. Salanitri, and B. Waterfield, *Future directions for the development of Virtual Reality within an automotive manufacturer.* Applied ergonomics, 2015.
25. Moudgil A (2019) AI & machine learning in automotive industries
26. Luckow A et al (2016) Deep learning in the automotive industry: applications and tools. In: 2016 IEEE international conference on Big Data (Big Data)

27. Sukitsch M, Engert S, Baumgartner RJ (2015) The implementation of corporate sustainability in the European automotive industry: an analysis of sustainability reports. Sustainability 7(9):11504–11531
28. Guenther H-O, Kannegiesser M, Autenrieb N (2014) The role of electric vehicles for supply chain sustainability in the automotive industry. J Clean Prod 90

Advances in Smart Maintenance for Sustainable Manufacturing in Industry 4.0

Abdelkrim Hadjadji, Sasan Sattarpanah Karganroudi, Noureddine Barka, and Said Echchakoui

1 Introduction

Generally, when it is necessary to collect data on the state of machines, we turn to technicians specialized in this area. In Maintenance 4.0, as presented in Fig. 1, with the rise of new connected technologies, these tasks can be performed by machines, which maximizes the useful life of machine components and avoids failures. With Maintenance 4.0 technologies, data meets humans and not the opposite. Maintenance processes evolve from a corrective and preventive model to a predictive one, changing the focus from diagnostic to prognostic. In fact, a key element of Maintenance 4.0 is predictive maintenance (PdM). This approach to monitoring machine health uses connected devices, thanks to the Internet of Things (IoT) embedded system, to collect data on a variety of assets. This approach delivers cost savings over routine or time-based preventive maintenance since tasks are performed only when predicted necessary.

Some of the advantages of this new digital era include monitoring the investment and return on equipment, overcoming communication boundaries, and projecting the organization onto the market. Big Data technology alongside Artificial Intelligence (AI), allows determining with higher precision the useful life of equipment, the risk of failure, the respective impact on the shop floor, and increasing the production reliability. As depicted in Fig. 2, the maintenance policy in industrial sectors governs the type of maintenance procedure deployment in the sector.

A. Hadjadji · N. Barka · S. Echchakoui
Département de Mathématique, Informatique et Génie, Université du Québec à Rimouski, Rimouski, Québec G56 3A1, Canada

S. Sattarpanah Karganroudi (✉)
Department of Mechanical Engineering, Université du Québec à Trois-Rivières, Drummondville, Quebec J2C 0R5, Canada
e-mail: sattarpa@uqtr.ca

H. Gholami et al. (eds.), *Sustainable Manufacturing in Industry 4.0*,
https://doi.org/10.1007/978-981-19-7218-8_6

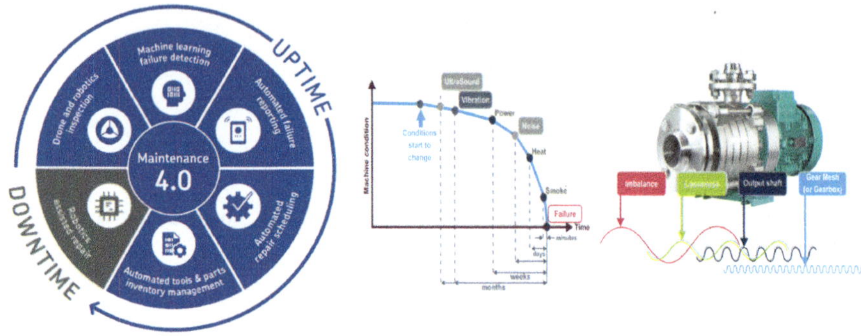

Fig. 1 The concept of maintenance 4.0

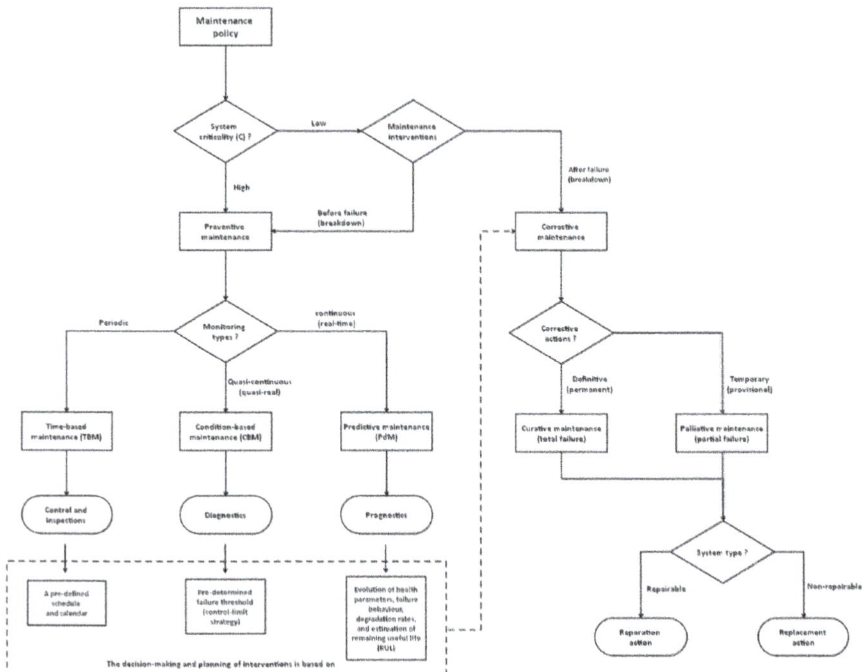

Fig. 2 Schematical flowchart of maintenance policy in industrial application

This chapter aims to cover the dissemination of original findings and new technologies in the planning, implementation, monitoring, and analysis of maintenance processes that support sustainable production in modern manufacturing companies. The remainder of this study is organized as follows. In Sect. 2, time-based and condition-based maintenance methods are described and compared. The predictive and preventive maintenance approaches and their utility in Industry 4.0 are then presented in Sect. 3. Prognostic and health management (PHM) as an engineering field applied in predictive maintenance, maintenance 4.0 tools such as cyber-physic systems (CPSs), Internet of things (IoT), Big data, and Artificial intelligence (AI) are introduced in Sects. 4 and 5. Digital twins, a virtual representation of equipment along with its simulated behavior, is described in Sect. 6 as an advantageous practice of smart maintenance in sustainable manufacturing which is capable of monitoring and assessing the remaining useful life (RUL) of assets in manufacturing sectors. Finally, a conclusive presentation of challenges and future directions in Maintenance 4.0 is presented in Sect. 7.

2 Time-Based and Condition-Based Maintenance

Historically, the notion of condition-based maintenance (CBM) and time-based maintenance (TBM), also known as systematic maintenance, first appeared in the 1970s in the context of preventive maintenance, responding to a particular interest of industries adopting a purely corrective maintenance policy "run-to-failure" resulting in an increase in damage and breakdowns that lead to unforeseen and unscheduled stoppages in production processes.

2.1 Time-Based Maintenance (TBM) «Systematic Maintenance»

Time-based maintenance (TBM) is a classical preventive maintenance approach adopted in various industrial fields based on the assumption that failure behavior or aging is predictable over time, more precisely, this assumption assumes that the failure rate (λ) of equipment during the life cycle always follows a bathtub curve divided into three phases: break-in or burn-in phase (decreasing λ), maturity phase (constant or quasi-constant λ), aging phase (increasing λ) [1]. Knowing that equipment failures truly related to aging (bathtub curve) represent only 15–20% of all failures, on the other hand, 80–85% are related to random conditions [2]. The objective of TBM is to reduce process losses and ensure production efficiency by improving equipment availability and performance and by reducing the rate of unscheduled failures and breakdowns [3]. Particularly applied to multi-component systems [4], it consists of the subsequent replacement of components likely to fail by available maintenance resources [5].

The most common systematic maintenance operations are component replacement, inspections or preventive visits, checks, adjustments, calibration actions, etc. [6]. These operations are carried out periodically according to predetermined schedules "fixed temporal periodicity" [7], or according to a periodicity of use (operating hours, number of units produced, etc.). TBM interventions are planned at fixed time intervals (expected lifetime) using information from the manufacturer's recommendations or analyses of the reliability and risk of failure (breaking down) of the equipment from information collected during operation [2, 6]. The process of applying TBM depends mainly on two steps. First, the analysis and modeling of the failure data collected on the equipment by statistical studies in order to estimate the mean time to failure (MTTF), and second, the decision-making on the intervention which is usually done based on the evaluation of the operational cost (failure cost and preventive actions cost) and also the evaluation of the type of the component (repairable or non-repairable). These decisions are based on an optimization approach, which allows to adjust and optimize the intervals of interventions [1].

The choice of TBM as a maintenance strategy is generally made because of its ease of management, decision-making, and execution, as well as the fact that the costs and charges associated with the intervention are known in advance and production stoppages can be negotiated beforehand. [6, 8]. Several problems arise from the application of TBM in industry. On the one hand, the most popular in terms of management and planning, the individualization or separation of the intervention periods of each component, leads to aberrant equipment-specific TBM planning. The solution to this problem lies in the ABAC-ABAD method [6]. On the other hand, in terms of profitability and efficiency, TBM still lags far behind other preventive maintenance strategies (conditional and predictive), as the inevitable waste of spare parts and periodic interventions on functioning equipment (healthy) add to the operating costs of the manufacturing processes and the cost price of the finished products.

Three obstacles also make it difficult or impossible to implement a maintenance plan based entirely on TBM: the technological peculiarities and specifications differ from one piece of equipment to another, and the recommendations and forecasts of machine constructors are not always reliable (several parameters, conditions and constraints influencing the real operation of the equipment after commissioning do not appear in the testing phase), the deliberate reduction of periodic maintenance intervals by machine constructors (to increase the replacement rate of spare parts in order to guarantee additional income) [9].

2.2 Condition-Based Maintenance (CBM)

Condition-Based Maintenance (CBM) is another form of preventive maintenance with a predictive doctrine applied in several fields such as renewable energy, industrial manufacturing processes, medical equipment, infrastructure and buildings, etc. Its main purpose is to optimize the efficiency and accuracy of maintenance decisions and activities within the preventive framework [10], more specifically, in comparison with TBM as shown in Fig. 3, it consists of quasi-continuously monitoring, assessing, and determining the health status of a piece of equipment in operation in order to carry out maintenance when needed [4, 10]. These actions will contribute on the one hand to decrease the life cycle cost of the equipment, and on the other hand to avoid catastrophic failures [1]. The decision-making approach in condition monitoring can be based on two methods: one based on diagnosis known as CCEB (current condition evaluation-based) and the other based on prognosis known as FCPB (future condition prediction-based) [1]. The current approach applied in most industries based on diagnosis is considered traditional or even obsolete, the latter consists of checking faults and performance status of systems or components [11], in order to provide warnings or early alerts on the abnormal operation of the equipment [1]. In contrast, a new approach based on prognostics, the latter allows the estimation and prediction of the approximate time before the equipment will be unable to perform its required function (will fail) [12]. The accuracy of this prediction reflects the performance of a conditional maintenance policy [13]. This estimation is based on failure prediction techniques and algorithms [14].

This maintenance strategy can be developed for real-time decision-making based on data processing by statistical approaches to estimate the remaining useful life (RUL) of the system [15] and to accurately set the date of intervention before failure, thus saving the costs of unscheduled maintenance and the costs of unnecessary periodic maintenance actions. Of the latter, the prognostic-based approach seems much more superior in terms of results than the diagnostic-based one [1, 10]. The application of such an approach can be translated into four approaches: data-driven, model-based, knowledge-based, and hybrid [3]. The implementation of a condition monitoring policy is mainly based on the monitoring, diagnosis, or prognosis of a multitude of parameters such as vibration, temperature, contaminants and impurities, and noise levels [1]. Monitoring and diagnosis are carried out on the basis of several of the most popular non-destructive testing (NDT) techniques: vibration analysis, acoustic analysis, oil analysis, or infrared thermography using sensors or measuring instruments attached directly to the equipment (on-line) [16], or by periodic sampling by the maintenance technician (off-line). In fact, this type of maintenance actually provides near-real-time diagnosis or detection of the current state rather than real-time predictions of the evolution of the degradation state of the equipment [16], in order to perform the necessary corrective operations when any of the state variables exceeds the predefined failure threshold [4, 17], also called the control-limit strategy [7, 18].

Nowadays, a major revolution in monitoring and diagnostic techniques and technologies and also data management [2], also called the scientific approach that combines statistics, mathematical programming, and artificial intelligence for rational decision-making through analysis of data collected in the process [1]. These advances have prompted the development of the maintenance function in general and conditional maintenance in particular, especially with the advent of the notion of industry 4.0 or the fourth industrial revolution as it was first known in 2011 at the Hannover Industrial Technology Fair [19].

As with any maintenance strategy, many limitations and factors affect the achievement of performance and efficiency of actions related to a condition-based maintenance (CBM) strategy, the most popular of which are: intervention planning time, imperfect condition information (masked failure effects), uncertain failure level (difficulties in accurately and precisely quantifying the failure) [13].

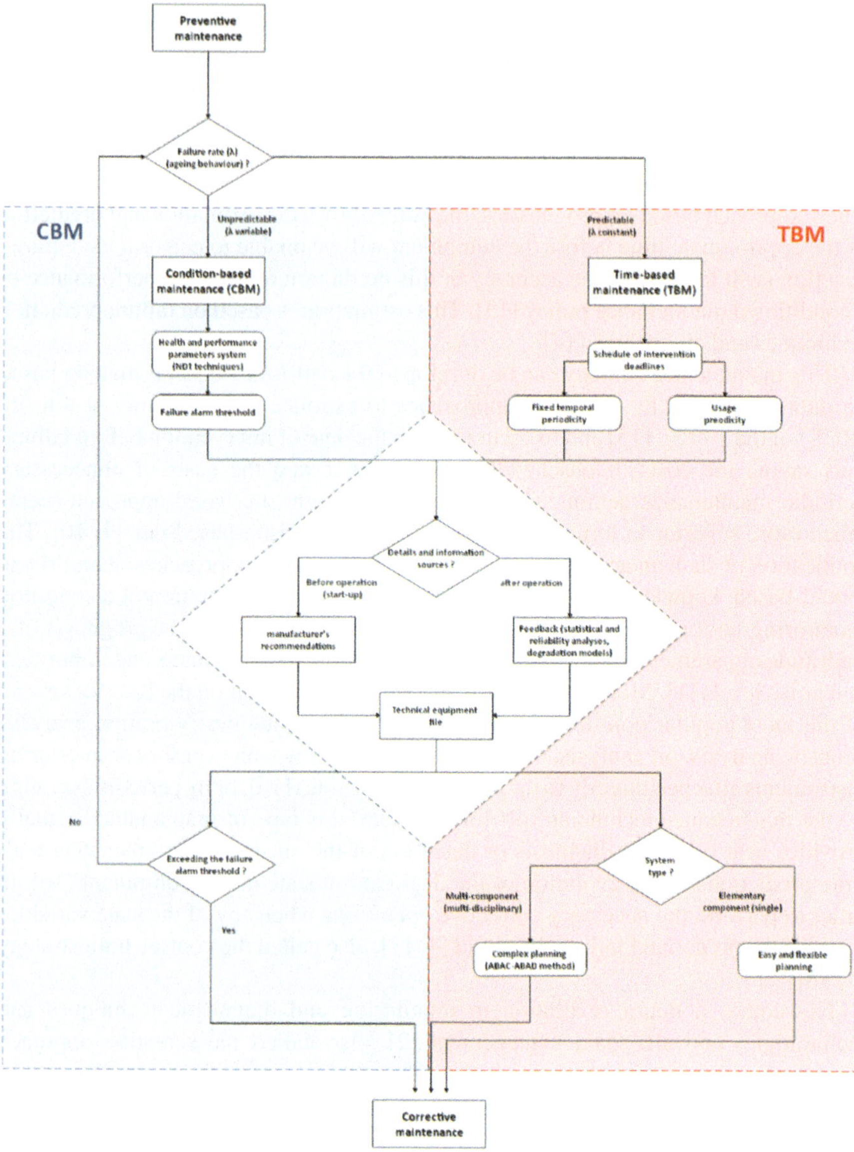

Fig. 3 Schematical comparison between condition-based maintenance (CBM) and time-based maintenance (TBM)

3 Preventive Versus Predictive Maintenance

Today, the manufacturing industry is faced with the obligation to adapt to the changes and challenges imposed by several economic, technological, and environmental factors. The latter has a major impact on the current situation of the industry, especially with the emergence of the notions of energy efficiency, green, renewable and sustainable energies, etc., triggered by the impact of climate change and greenhouse gas emissions (CO_2). The present conditions impose the transition toward a sustainable industrial model based on the technological development of tools and techniques offering new opportunities in the era of Industry 4.0. In the light of this transition, and like any industrial activity, maintenance is chasing time to make the transition to predictive maintenance or maintenance 4.0, especially for classical preventive approaches.

3.1 Preventive Maintenance (PM)

Preventive maintenance is more than a maintenance approach or strategy adopted by the manufacturing industries. It is in fact a whole line of thinking that emerged in the 1970s in response to the traditional 'walk to failure' practices used in corrective maintenance. This reasoning includes approaches that are primarily aimed at eliminating the notion and phenomenon of fortuitous and unforeseeable failure or breakdown. Preventive maintenance is a set of actions aimed at preventing failures, reducing the risk of failure, and the number and time of unscheduled downtime, which implies an extension of the equipment's life, it consists of intervening in a system before a failure occurs. According to [20], the transition to preventive maintenance provides the advantage of saving more than 18% of maintenance costs for companies adopting a purely corrective policy. In general, the preventive approach has two main categories: systematic maintenance and condition-based maintenance [20–24].

The two approaches of systematic and conditional preventive maintenance are already developed in the previous section. Overall, for systematic maintenance, the principle is based on the planning of maintenance activities periodically at fixed intervals in time or even according to prescribed criteria of the units of use of the machine, these intervals are determined from the average life of the equipment based on historical data, experience feedback, and manufacturer's recommendations. According to [20, 23], this approach is very costly for 92% of the equipment parts and components. While for condition monitoring, the principle is based on monitoring the operating condition and detecting the faulty behavior of the equipment in order to establish corrective maintenance actions at a predetermined threshold with the help of data analysis based on statistical approaches, manufacturer's recommendations, feedback, etc. According to [21], there are three types of monitoring in CBM: On-demand monitoring, scheduled monitoring based on inspection, and continuous monitoring based on the use of sensors. It is true that preventive maintenance does not represent the best optimal choice in relation to the maintenance cost, time and number of stops, production rate and key performance indicators (KPIs), the maintenance policy to be adopted, but it presents the best alternative to purely corrective maintenance for industries [20].

3.2 *Predictive Maintenance (PdM)*

The implementation of a predictive maintenance approach within the framework of Industry 4.0, also known as smart industry, gives manufacturing companies the possibility of reducing total downtime by 30 to 50%, increasing equipment life by 20 to 40% [19, 25, 26], achieving savings of 30 to 40% [27], decrease the number of unnecessary or unplanned downtime during the equipment life cycle and also optimize costs (energy costs by minimizing energy consumption and decreasing labor costs and spare parts consumption), control and production quality [21, 26, 28]. More specifically, according to [20, 22, 23], a PdM plan leads to a tenfold return on investment, with a 35–45% decrease in downtime accompanied by a good troubleshooting margin. Therefore, the elimination of breakdowns of 70 to 75% implies a 20–25% increase in production and a 25–30% decrease in maintenance costs, where according to [29] maintenance costs can contribute 15–70% of the total cost of production (cost price). While achieving such results and success implies profound changes in the culture and practices used, technological improvements and significant investments in the manufacturing process (in many cases entailing more than 30,000 € per equipment) [20, 21], and also the adaptation of the recruitment and human resources management policy with the obligations, requirements, and advances of the field, where PdM requires the recruitment of reliability and data science engineers and the training of staff [20].

PdM is a preventive maintenance approach that aims to improve the performance and efficiency of the manufacturing process by increasing the lifetime of equipment, which implies on the one hand a decrease in downtime and the number of unnecessary shutdowns accompanied by a reduction in direct and indirect costs related to maintenance. According to [21], in the United States, almost 33% of the maintenance budget is spent on unnecessary actions. The result is very high reliability, availability, and productivity rates by improving the safety of people and equipment (a very low-risk level) [21, 23, 30]. Predictive maintenance is nowadays one of the most developed maintenance policies. In fact, it is a sophisticated version of condition-based maintenance, which is based on continuous (real-time) monitoring of the equipment's operating state, It is not limited to the phase of detecting signs of failure and locating the fault only (breaking with other approaches), but it aims to anticipate impending failures, predict the faulty behavior of the system, and also reliably estimate the remaining useful life (RUL). These actions build the core of the prognostic approach [21, 31]. This approach, closely linked with PdM, is usually based either on data or a model [21]. The difference between these two approaches is not easy to determine, it is somewhat blurred, where the data-based approach consists of establishing analyses and interpretations on historical system data, while the model-based approach is applied to systems that do not have historical data [21]. According to [22, 32], PdM is divided into two categories depending on the failure detection methods: one based on statistics and the other on conditions.

The prognosis is not only limited to estimating the remaining useful life (RUL), but can also be used to assess the energy efficiency of the equipment and the environmental effects associated with the failure [26]. This approach provides a detailed examination of data from feedback, manufacturer's equipment knowledge and recommendations, key performance indicators (KPI), and information collected and gathered in the terrain by multiple well-developed sensors installed in the manufacturing process [24, 28, 31, 33], these data must be available with a sufficient and representative quantity to ensure the effectiveness of the PdM approach [31]. The analysis and interpretation of these data are usually done using statistical inference models, regression models, machine learning models, etc. [34].

The process of applying a predictive maintenance approach according to [35] is based on five modules: sensor selection and data acquisition module, data preprocessing module, data mining module, the decision support module, and maintenance action module, however, according to [22], it is based on three steps: data acquisition, data processing, and maintenance action decision. The prediction approach adopted in the PdM concept can be based on different models: based on a physical model through mathematical modeling of the health state of equipment, or based on the knowledge characteristic to the system, or based on data-driven steering based on statistical models or artificial intelligence and machine learning or deep learning models [28], this prediction offers the advantage of planning the maintenance intervention in optimal time before the system deteriorates or fails [35].

The decision-making phase in a predictive maintenance strategy, which is gaining in accuracy and becomes more and more efficient in handling complex data, especially with the advent of the smart industry era and the technological development of emerging tools and methods in maintenance such as cyber-physical systems, Internet of Things (IoT), big data, data mining, artificial intelligence and machine learning (ML), etc. [36]. This decision efficiency essentially helps to optimize maintenance activities and minimize the catastrophic impact of an unexpected failure. The predictive maintenance strategy is also known under the concept of e-maintenance when all the equipment constituting the manufacturing process are interconnected, this concept represents a broader and more comprehensive vision of the next generation of the manufacturing industry that gives many benefits and addresses the basic need of artificial intelligence tools in relation to the maintenance activity in manufacturing companies [37], this PdM integration contributes to reducing maintenance costs by 10–40% [22, 25, 26]. It can also be adopted in a concept of outsourcing of maintenance activities, i.e., a whole maintenance policy purely outsourced by leaders and experts of the field, which will help to increase the effectiveness, performance, and efficiency of maintenance operations, and make manufacturing companies more powerful in terms of competition and competitiveness [23].

Like any industrial activity, sustainable predictive maintenance faces many threats, risks, and obstacles. The most common ones are financial, environmental, and social obstacles, but we can also add other obstacles such as: organizational, technological, and also depending on the nature, type, and technological level of the industries [38]. The predictive maintenance strategy still remains a real challenge despite the explosion of computerization and digitalization notions in the last decade, it is, on the one hand, an important technological development in tools, techniques, and even in promising ideas, but on the other hand, a difficult reality on the ground and a situation with many obstacles and constraints risking the success of this strategy. More specifically, the weak commitment and resistance of manufacturing companies adopting traditional maintenance policies to make the transition to PdM, especially with the lack of investment and financial resources related to the global economic crisis caused by the pandemic COVID19, which now represents a real problem that can firstly hinder progress and research in the field, and secondly delay the implementation of PdM and the feedback of experience.

In reality, PdM is not and will never be an alternative to preventive maintenance, but just a natural extension and inevitable development of the preventive policy over time under the influence of changes, technological improvements, challenges, conditions, and variations in the field. Therefore, the aggregation and presence of all these ingredients at the same time is the foundation of this current version of PdM. For example, if one of these elements was missing or another element occurred, it would never be the same version, nor the same principles and nor the same ambitions of the PdM as today.

4 Prognostics and Health Management (PHM)

4.1 PHM Concept

PHM is a vision or discipline adopted by manufacturing companies in the context of maintenance activity for the assessment of the actual condition based on a comprehensive set of methods, techniques, and technologies for the monitoring of the overall health, diagnosis, and prognosis of equipment to determine reliability, accurately estimate the remaining useful life (RUL bases as presented in Fig. 4) or estimated time to failure (ETTF) with minimum uncertainties, and predict the future state of potential failures and decrease the occurrence of unscheduled failures [1, 39–41], with the aim of optimizing the life cycle and ensuring the efficiency of maintenance operations [19, 42, 43]. The adoption of such a discipline makes the manufacturing system intelligent [44]. As a principle, PHM before being applied in maintenance was initially introduced in the field of medicine for the early detection of diseases [45].

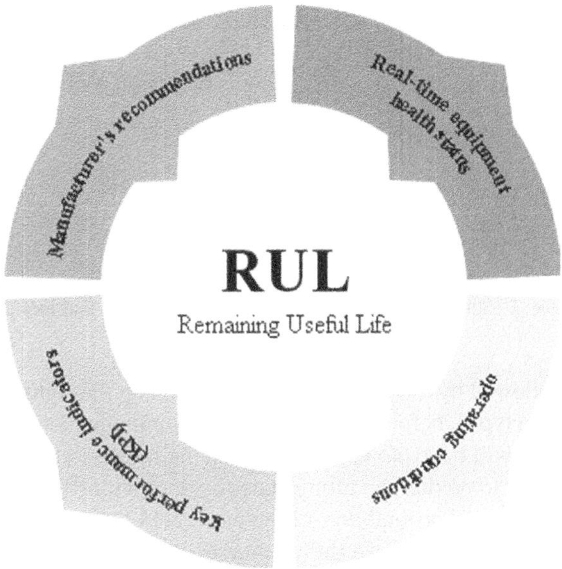

Fig. 4 The bases of remaining useful life (RUL) approach

The PHM process often consists of observation, analysis, and action mainly using sensors, algorithms, and mathematical and machine learning models for the estimation of RUL, performance, and reliability of products, subsystems, and processes under real-life conditions [46]. PHM is a systematic approach that also considers the energy consumption of processes for decision-making in the context of sustainable manufacturing using performance indicators such as the Energy Efficiency Indicator (EEI) [45].

Guillén et al. [40] and Atamuradov et al. [47] proposes a PHM process model designed by ISO 13374/OSA-CBM that contains three phases: detection and conditioning, diagnosis, and prognosis described in ISO 133811. This PHM model in the e-maintenance concept gives manufacturing companies the advantage of being conscientious and competitive. In a general view, as presented in Fig. 5, a PHM system model comprises several steps under four main headings: process preparation steps, data management process, PHM milestones, and finally the adoption of PHM results, however, an effective PHM system model requires three key steps: estimating and determining the current health state, predicting the degradation behavior, future state and remaining life (RUL), estimating and evaluating the extent of damage and the impact on system performance [41, 47].

4.2 PHM for the Manufacturing Industry

The implementation of a PHM discipline within manufacturing processes requires first a reliability assessment and criticality analysis to prioritize critical components,

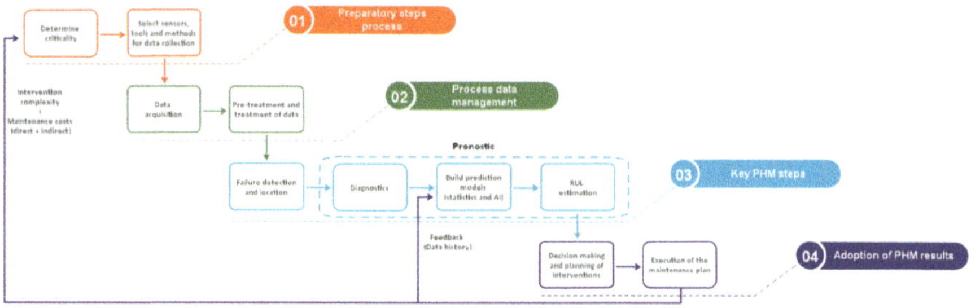

Fig. 5 Schematical descriptive view of Prognostics and Health Management (PHM)

followed by a cost analysis, then choosing the right sensor types for the system and data types, before the adaptation and implementation of the collection process. This data will be exploited afterward in the framework of diagnosis and also prognosis in order to predict the future state together with RUL and to improve the performance, safety, and profitability of the process [42, 44, 47]. This profitability is often linked to the ratio between the volume of initial investments and the benefits obtained by the application of the PHM.

According to [47]**,** the implementation of a PHM discipline consists of seven key steps: data acquisition, data pre-processing and processing, detection and localization, diagnosis, prognosis, maintenance decision- making, and execution from the human–machine interface. The application of such a discipline often leads to radical changes and modifications in the plans, practices and policies of production, maintenance, logistics, etc. [40]. For maintenance, PHM's main objective is to advance and expand the use of preventive techniques, particularly predictive techniques, at the expense of corrective techniques [43]. A PHM project (PHM4SMS) for intelligent manufacturing systems was launched by the NIST (National Institute of Standards and Technology) in 2013 [42]. This organization has gathered all the key needs and priorities of manufacturing companies in terms of PHM strategies and maintenance policies [43]:

- Providing the means, techniques, and technological tools necessary for the success of the PHM process (monitoring and detection, diagnosis, and prognosis).
- Implementing and preparing the data management process in order to provide real data to achieve optimal efficiency and accuracy of diagnosis and prognosis.
- Determining key performance indicators from the different surveillance techniques.
- Promote and encourage the exchange of PHM knowledge and experience between the manufacturing community.

The use of this discipline of PHM in manufacturing industries remains very limited, almost rare, where their principles still seem very complex and difficult to apply because of several factors despite the huge technological development of tools and techniques caused by the arrival of the fourth industrial revolution "industry 4.0". These factors, on the one hand, are of a financial nature (lack of investments especially for small and medium-sized enterprises (SMEs)), social (lack of expertise and qualification of staff), technological (data collection and processing), etc. On the other hand, they depend directly on the complexity of the manufacturing processes which consist of various interacting equipment with different designs and architecture, different failure modes and degradation rates, different and interdependent operations and tasks, different data types, and also RUL different, the thing that complicates the decision- making in terms of prognosis and adopted maintenance policies and that implies multiple PHM techniques and tools [42, 45, 48, 49]. Meanwhile, the application of the PHM approaches separately for elementary and typical components by using prognostic techniques such as vibration analysis, infrared thermography, oil analysis, ultrasonic analysis, etc., proves to be very reliable and efficient [48]. According to [45, 49], the main challenges for the implementation of a PHM approach for manufacturing processes are:

- Equipment-specific failure modes and rates.
- Complex and distinct equipment architecture and design.
- The many peculiarities and details of production operations.
- Difficulties and complexity in terms of scheduling.
- The need for comprehensive prognostic tools and techniques capable of studying and processing data from the entire process accurately and efficiently, i.e., studying and processing data from several failure modes, parameters, components, and subsystems simultaneously (on all levels).
- The technological tools and techniques used require a significant investment in the manufacturing structure.
- In return, there is a lack of resources for the research and development of PHM models and practices, especially for SMEs [44].

5 Maintenance 4.0 Tools

At the beginning of the twenty-first century, and in the era of the fourth industrial revolution, recent technologies such as cyber-physical systems (CPS), internet of things (IoT), internet of systems (IoS), Big Data, etc., have been converted in the industry as main components of digital transformation in the framework of Industry 4.0. These have been adopted by manufacturing companies in multiple industrial applications such as maintenance, more specifically predictive maintenance or also maintenance 4.0, specifically addressing issues related to maintenance decision-making in direct relation to big data analysis and interpretation, failure prediction, and remaining life estimation (RUL). In this regard, they can position themselves and stand up to the fierce economical competition.

5.1 Cyber-Physical Systems (CPS)

Cyber-physical systems represent the latest technology of intelligently connected production devices, sensors, and autonomous intelligent monitoring, control, sensing and diagnostic systems equipped with a set of computational and physical techniques and tools fully aligned with human needs [35, 50]. The concept of CPS was first introduced in 2006 by the American scientist Hellen Gil at the National Science Foundation (NSF) in the USA around 2006 [28, 51]. The activity of CPS is decentralized and self-organized and they are characterized by homogeneity and flexibility in working with each other in different disciplines and modes of operation [52, 53]. This key element of Industry 4.0 has several automated sensors (mechatronic components) and is distributed across all equipment for data collection, storage, and analysis. CPSs represent the link between the virtual and the physical world, more exactly, they give the possibility to merge the virtual world with the real industrial world by eliminating all boundaries between them, and they also allow decentralized decision-making [35, 51, 53–55]. These systems provide continuous process monitoring and data exchange through a virtual space (IoT tools) and allow remote diagnosis and effective decision-making in terms of real-time maintenance, thus contributing to the creation of intelligent manufacturing processes [51, 52, 56, 57].

The contact and interaction of CPSs with humans are usually done through Human Machine Interfaces (HMIs) [52]. In the industrial sector, especially in the context of predictive maintenance or maintenance 4.0, the use of this technology is often known as CPPS (Cyber-Physical Production Systems) [58, 59]. Apart from the manufacturing industry, CPS technology is widely used in a multitude of fields such as aerospace, automotive, energy, health, and transportation [53].

5.2 Internet of Things (IoT)

IoT is considered as the main factor in the emergence of the Industry 4.0 principle (intelligent industry) and as the basic technology of cyber-physical systems. Simply, it is the network linking the CPS ensuring the interconnection and interaction between the physical devices of the process, which allows the automatic collection and retrieval of the huge flow of data generated by the various devices of the process constituting the notion of Big Data. The IoT also provides the possibility of data transmission via the internet, and it is the atmosphere that allows direct access to process data, virtualization of resources, interconnection, cooperation, intercommunication, and machine-to-machine (M2M) interaction in a transparent way and without human intervention [35, 51, 55–58]. The adoption of this internet-based technology by manufacturing companies helps to improve intercommunication and interaction between machines, increase the performance and quality of maintenance operations, and avoid unexpected failures, as well as resource optimization and cost reduction [55].

The concept of IoT is widely spread in various fields, and the most popular different IoT technologies are Extensible Messaging and Presence Protocol (XMPP), Data Distribution Service (DDS), Advanced Message Queuing Protocol (AMQP), Message Queue Telemetry Transport (MQTT), Open Platform Communications-Unified Architecture (OPC-UA). The latter is the most practiced and used by large manufacturing industries [56]. According to [53], the design of IoT systems is in the form of three layers (multi-layer system), namely: IoT platform layer, IoT application layer, and IoT industrial solutions layer. This design provides several advantages in terms of process efficiency and flexibility. In the industrial sector, especially in the context of predictive maintenance or maintenance 4.0, the use of this technology is often known as IIoT (Industrial Internet of Things) [28, 51, 58, 60]. According to [60], in a survey of a set of manufacturing companies on the adoption of this technology as part of a predictive maintenance policy, almost 44% of companies are already applying it, and 27% are planning to use it in the near future.

5.3 Big Data

The collection, processing, and analysis of real-time Big Data from CPS is a strategic phase for the intelligent transformation of the maintenance function, especially for PdM in relation to failure prediction, planning, and optimization of interventions, and is mainly done by artificial intelligence tools and techniques such as machine learning (ML) and deep learning (DL) or by models and statistical approaches based on data and information collected by CPS [55]. The objectives of this approach are the implementation of models and algorithms for failure prediction, continuous process control and monitoring, and failure diagnosis and prognosis.

The concept of Big Data is mainly characterized by four parameters such as the volume of data, the variety of data, the speed of analysis and production of new data, and also the quality and value of data [55]. Big Data in the manufacturing industry is synonymous with huge, diverse, complex data of different volumes and nature from several sources reflecting the actual state of the process impossible to analyze and process by conventional techniques and tools. These data are usually a mixture of structured, unstructured, and semi-structured data [36]. According to [36], types of data can be classified into Big Data, Structured, unstructured, and semi-structured data, Time-stamped data, Historical data, Operational data, Identity data, Asset data, and Environmental data. These different types of data originate from sources such as equipment design, machine operation, equipment health conditions (conditional maintenance), periodic interventions (systematic or corrective maintenance), deliverable quality, process logistics, value chain costs, customer requirements and conditions (feedback), operator behavior, and environmental conditions [36].

5.4 Artificial Intelligence (AI)

Artificial intelligence is the keyword in the transition to Industry 4.0, it is a powerful technology that compensates for the deficiencies and ineptitude of traditional techniques and approaches practiced in industry. AI is strongly linked, appropriate, specific, and exclusive to Big Data in order to answer critical questions, remedy weaknesses and shed light on key process issues, specifically in the analysis and processing phase of big data. The latter is done mainly through the use of techniques such as machine learning (ML) on data collected from different sources to detect the abnormal operation of equipment, extract and guess the degradation patterns of components, know the relationships between different parameters and components, strengthen the decision-making in terms of maintenance and failure prediction and process optimization, where it offers manufacturing companies several economic benefits on finance, production, logistics, maintenance, etc. Especially in the predictive aspect, these benefits are gained thanks to AI in effectiveness and prognostic accuracy in terms of failure prediction and remaining life estimate (RUL), in speed and quality of maintenance interventions by reducing the error rate, in planning and efficiency of maintenance plans, in profitability and optimization in terms of resources and investments.

5.5 Connection Interfaces

Represent spaces and platforms for communication, sharing of actual equipment status, information exchange, and storage of useful data linking different disciplines, partners, customers, and suppliers in the process. They offer the advantage of remote expertise in the context of Maintenance 4.0 for complicated maintenance operations, or unusual failures and breakdowns. In several disciplines or domains, these interfaces are the means of communication and interaction with humans, such as Human Machine Interfaces (HMI). Nowadays, all these key components present the foundation of Industry 4.0 such as CPS, IIoT, Big Data, and connected interfaces like the Cloud and the cooperation, coordination and interaction between them and the human, play a key role in the industrial transition in particular and the economic transition in general toward the sustainable context or the factory of the future (FoF) and smart manufacturing. These technologies in the context of Maintenance 4.0 help on the one hand to improve the performance of equipment, the quality of products, and the efficiency and profitability of manufacturing processes, and on the other hand, to overcome the difficulties and limitations related to the classical tools and techniques used in industry. In other words, these technologies strongly contribute to the transformation of manufacturing companies and the flexibility of operations by creating an intelligent industrial atmosphere in the framework of the orientation toward Industry 4.0, the driving force of a sustainable industry. Virtual reality can be used in this sense to facilitate man–machine contact, improve the quality of interventions and help operators to take optimal solutions.

The challenge today is to make these key technologies of Industry 4.0 (CPS, IIoT, Big Data, etc.) autonomous and capable of interacting with other objects, devices, and external systems from different domains and disciplines. This challenge extends to make real-time information exchanges using connected and common interfaces without human intervention [56, 57] and also to protecting and securing the data and information shared through these components which are not immune to cyber threats.

6 Digital Twins and Real-Time Monitoring, Tools of Efficient Production Management

The emergence of new technological and innovative trends in the era of the fourth industrial revolution such as CPS, IoT, Big Data, AI, and Digital Twins (DT) has prompted the development of new production methods addressing key issues for the transition to sustainable and smart manufacturing. Digital Twins in this context of Industry 4.0 and despite the significant delay in aspects related to DT standards, the formal framework of the DT concept, and safety, considered as one of the robust and promising technologies offering advantages of efficiency, competitiveness, and profitability, especially for the manufacturing industries.

6.1 Digital Twin's History

The appearance of the concept of digital twins dates back to the beginning of the twenty- firstcentury with the emergence of simulation as a powerful tool in the scientific as well as the industrial world, exactly, this technological trend was first introduced by Michael Grieves in 2002–2003 as part of his course on Product Lifecycle Management (PLM) [61–66]. Afterward, this concept was presented to the general public by NASA (National Aeronautics and Space Administration) in 2012 within its technology roadmaps [61–63, 67]. Today, DT technology is increasingly recognized and attracting interest from researchers in academia and also from manufacturing companies, where it has been considered according to Gartner in three consecutive years (2017, 2018, 2019) as one of the ten most promising and strategic technology trends in the world [61, 62, 68].

6.2 Digital Twins Concept

DTs play a key role in virtualizing and digitizing the main complex physical objects, functions, and systems constituting manufacturing processes, they represent the global virtual version, copy or model of the real and multidisciplinary physical world (production process) that encompasses and integrates the interacting set of all key physical components (products, resources, orders), their data, monitoring, control, planning and scheduling tools, and the history of any process [58, 68], in order to improve the productivity and efficiency of production operations, increase process performance and flexibility, and continuously optimize production management, which strongly contributes to the achievement of significant cost savings and reduction of costs and expenses during the process life cycle [62, 63, 65]. In other words, DT is seen as the manifestation of physical entities in the digital (virtual) domain [66].

A DT system collects all data, information, and details related to the physical entity during its life cycle, from the design phase to the decommissioning or end-of-service phase. This system enables real-time condition monitoring and control, entity modeling, continuous behavior prediction and real-time decision support, and optimization of the performance of physical entities in the real world in order to maximize the profitability of the production process and improve operating times [61, 63, 65, 66, 68]. According to the IDC (International Data Corporation), a 30% improvement in cycle times is expected in five years for companies adopting this technology [61, 66]. DT can be summarized as seamless integration between virtual space and the real physical world, which provides and feeds manufacturing processes with data and information on production management, actual equipment status, and potential failure behavior globally promoting process intelligence and flexibility with the aim of consolidating the key principles of sustainable and smart industry [62, 69].

DTs can be divided into two pillars as virtual DTs of the physical entities of the manufacturing process, and predictive and decisional DTs based on physics-based, data-based, or hybrid models and which have the main objective of event prediction and process optimization [58]. In some contexts [68], TDs can also be decomposed into three pillars with the addition of the projection of DTs through the integration and adoption of the results generated by predictive DTs into the manufacturing process. The design of DTs includes three dimensions: physical part, virtual part, and the connection between the two [63, 69]. While in some works of literature, the complete design of DTs includes five dimensions by adding data and services [62, 63].

6.3 Advantages of Digital Twins s for Manufacturing Industries

According to [68], the main benefits of the DT concept are as follows:

- Real-time status monitoring and remote control.

- Excellent efficiency, accuracy, and safety of the operations performed.
- Prediction of failures with optimal accuracy and optimization of maintenance planning (PdM).
- Study of system behavior and risk assessment in virtual space to minimize losses and eliminate the danger.
- Better collaboration, consistency, and interaction within and between teams.
- Efficient, fast, and rational decision support process.
- Faster and smoother adaptation to market changes and customization of products and services.
- Facilitates communication, interaction, and real-time exchanges by improving documentation.

The concept of DTs is a recent technology that has not been widely adopted by manufacturing companies around the world. The applications of DTs within the manufacturing industries are mainly concentrated in the fields of:

- Design, especially for the manufacture of complicated and costly prototypes (aeronautics and aerospace).
- Production mainly for condition monitoring (KPIs), resource, and supply chain management.
- Predictive maintenance, especially in prognostics and process health management (PHM) for real-time monitoring, prediction of remaining equipment life (RUL), and also for decision-making to optimize intervention planning [62, 63, 68].

DT technology still faces many challenges in the manufacturing domain mainly related to safety, data quality, accuracy and correctness of failure predictions and remaining life (RUL) (optimizing tolerance intervals and minimizing uncertainties), real-time simulations, Big data management, and fast and rational decision-making for the development of the smart manufacturing concept [68].

7 Manufacturing and Maintenance at the Heart of Sustainability

7.1 Sustainable Manufacturing

The concept of sustainability first appeared in the 1980s within the manufacturing industry under the name "cleaner production", focusing essentially on production as an axial function in the industrial sector with the objective of improving the degree of exploitation of raw materials by minimizing the resulting losses and rejects along the manufacturing cycle [70]. This vision of sustainability is fully consistent with the Japanese "Muda" approach based on Value Stream Mapping (VSM) within the framework of the Lean philosophy [71].

According to [72], the idea of sustainable development was first introduced in 1987 in the Brundtland Report at the World Commission on Environment and Development.

The concept of sustainability is a multidisciplinary field based on three fundamental dimensions: social, economic, and environmental [73–76], also known as performance indicators according to the Global Reporting Initiative (GRI) [72].

In the era of Industry 4.0, the adoption of new technologies (CPS, IoT, Big Data, IoS…etc.) as part of the digital transformation of the manufacturing industry has significantly influenced the sustainability of core manufacturing activities (production, maintenance, logistics…etc.) under the requirements of the various dimensions of sustainability, including process and production line performance, productivity, and profitability of the maintenance activity and reduction of wasted expenditure, energy efficiency, human reliability, reduction of environmental impact…etc. [74].

According to [73], the sustainable manufacturing approach is mainly based on four strategic pillars covering all the concerns of the industrial world, the stages related to the life cycle of the product from the reception of the raw material to the delivery of the finished product, contributing significantly to achieving the global objectives of sustainability in terms of efficiency, profitability. These pillars are: waste reduction, material efficiency, resource efficiency, and eco-efficiency.

Today, sustainability is attracting the attention of researchers in the academic world and also manufacturing companies, several studies and a variety of methods have been developed with the aim of achieving sustainable manufacturing, [71] have reviewed in the literature the use of sustainability-oriented VSM also called "Sustainable VSM" in manufacturing processes by proposing both a set of sustainability indicators to identify the effect on economic, social, and environmental sustainability.

The reduction of waste and consumption of raw materials is only part of the solution, there are other solutions that can be relied upon in the context of sustainability, including increasing the functional life of products, preserving raw materials, and rationalizing technological development to make it suitable for current and future needs.

According to [76], the adoption of the "6R" concept (reduce, reuse, recycle, recover, redesign and remanufacture) within manufacturing processes can support the achievement of the goal of sustainable manufacturing.

7.2 *Sustainable Maintenance as Part of Sustainable Manufacturing*

The integration of sustainability concepts into most of the key activities of the manufacturing processes can only be achieved with the inclusion of sustainable maintenance, which is one of the main foundations of sustainable manufacturing aiming primarily at maintaining the balance between the different dimensions of sustainability, optimizing the life cycle management of assets, increasing the life and utilization rate of machines, minimizing the consumption of materials and non-renewable energy (conventional and fossil fuels), reducing pollution and improving environmental conditions, reducing costs, improving equipment performance and system productivity [76–79].

Sustainable maintenance is the typical version of maintenance 4.0 ensuring short, medium, and long-term sustainability (economic, social, and environmental), value creation, satisfaction of needs, availability and performance of the industrial process for future generations [80]. Indeed, sustainable maintenance is a collective version merging all existing maintenance policies from corrective to predictive or M4.0 in a way that helps to extract the best features and properties of each policy in order to achieve the objectives of sustainability.

In contrast to the classification of [72–76] of the sustainability dimensions in sustainable manufacturing (economic, environmental, social), [77, 78] have divided sustainable maintenance according to its significant impact into four dimensions: technical, economic, environmental, social and safety.

In line with the current trend of Industry 4. 0 and the adoption of its technological tools at the heart of the maintenance activity, [81] the software publisher Dassault Systems offers Manufacturing Operations Management (MOM) with the DELMIA application and the 3DEXPERIENCE platform, as well as Manufacturing Execution Systems (MES), which enables manufacturing process efficiency and resilience, continuous and remote equipment monitoring, control and diagnostics, maintenance planning and management, and the use of artificial intelligence for decision support. As shown in Fig. 6, the overall concept of sustainable maintenance is in the context of sustainable manufacturing.

In their report, the European Commission [82] identified nine (9) critical success factors (CSFs) that are essential in managing maintenance activities in small and medium-sized enterprises (SMEs) in order to implement a sustainable and lean maintenance approach.

The major challenge facing manufacturing companies today in implementing a sustainable maintenance policy is to overcome the difficulties and constraints related

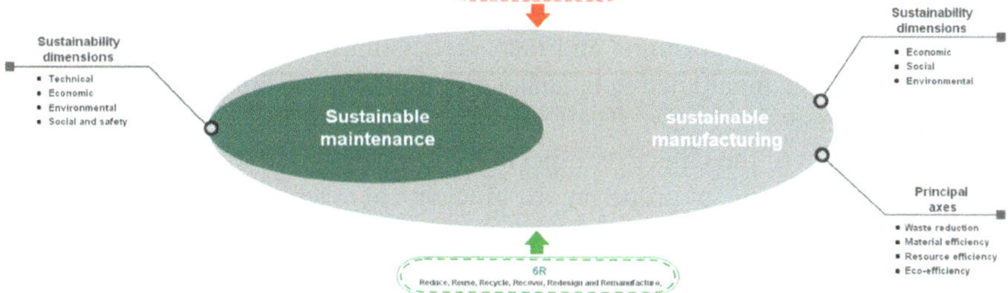

Fig. 6 The concept of sustainable manufacturing and maintenance

to the complexity of manufacturing operations, the fierce competition in the economic market, and especially those related to the sudden change in practices and habits of previous policies that will undoubtedly cause destabilization in the manufacturing process.

8 Challenges and Future Directions

The concepts of maintenance 4.0, innovative technologies, and artificial intelligence correspond today to a new way of organizing the means of production. This new form of industry, which holds great promise for consumers, is the convergence of the digital world and financial and economic operations with the products that exist in our reality. It should be noted that Maintenance 4.0 touches multiple aspects such as economic, social, political, and environmental. The priority of adopting Maintenance 4.0 in maintenance routines of manufacturing sectors is recognized by industrial companies. In terms of challenges and future directions, there are a number of measures to be taken into account in order to be able to challenge this novelty. To this end, it would be interesting to list in summary form the possible actions that we have retained in this perspective:

(1) Study and exploration of new prognostic techniques and approaches capable of:

   - Simultaneously monitoring multidisciplinary systems
   - Monitoring the various parameters and factors of dysfunction
   - Simultaneously studying all the different failure modes and the interaction between them and the influence of one on the other
   - Predict the different degradation behaviors
   - To estimate the remaining life (RUL) of the system, knowing that most of the prognostic techniques and approaches proposed in the previous research are limited to the study of a single failure factor or mode, these techniques are also unable to estimate the remaining life (RUL) characteristic of the complete system.

(2) Proposal of tools and techniques to improve the flexibility of predictive maintenance under different conditions and situations in order to optimize and reduce the planning and preparation time of interventions.

(3) Development of algorithms and data processing tools capable of accurately processing and manipulating large heterogeneous data from different sources at the same time.

(4) Achievement of the objective, especially with the economic crisis linked to the Covid19 pandemic, of preparing the efficient approach and proposing the ideal path for the transition and transformation phase at a lower cost for industries adopting traditional corrective or preventive maintenance policies. Therefore, do not just implement the maintenance 4.0 policy with its key components for the digital and intelligent transformation in the framework of Industry 4.0.

(5) For the manufacturing industry, currently, the challenge is to define the general framework (the core), the environment, and the steps for the application or implementation of a maintenance 4.0 policy or predictive maintenance appropriate to all manufacturing sectors, so that each company can adjust it according to the manufacturing process it has, while respecting the requirements, needs, details, characteristics, and particularities of each equipment, operation, stage, area of manufacturing, etc.
(6) Development and strengthening of cybersecurity of manufacturing processes adopting those key technologies to the digital transformation such as CPS, IoT, connected interfaces, Cloud, Big data, etc. These provisions are to be recommended against cyber risks and attacks. It should also be noted that these technologies have not yet proven to be effective in protecting data. In addition, each manufacturing process has a wide variety of highly confidential data and information.
(7) Increasing and improving the skills of the personnel that are essential for the successful transition to the smart and sustainable factory in order to be able to keep pace with the technological development of industrial processes.
(8) In-depth study of the impacts, advantages, and disadvantages caused by the adoption and use of these technologies in the industrial sector, in particular for the manufacturing industry, in order to rationalize investments in an equitable way ensuring the sustainability of the key elements of the manufacturing processes including the human being, as the main aim of this technological development in the era of industry 4.0 era is to facilitate work and optimize human performance, not to eliminate or exclude it altogether (the self-congratulatory benefit provided by these technologies will inevitably contribute to rising unemployment rates, negatively influencing economic recovery).
(9) Careful study of the economic impact, i.e., the profitability of adopting maintenance 4.0 as predictive maintenance. This can be done independently or linked to a whole Industry 4.0 program by manufacturing companies in the short and medium-term, especially SMEs (lack of financial resources, lack of research and development teams).

References

1. Ahmad R, Kamaruddin S (2012) An overview of time-based and condition-based maintenance in industrial application. Comput Ind Eng 63(1):135–149. https://doi.org/10.1016/j.cie.2012.02.002
2. Zou G, Banisoleiman K, González A, Faber MH (2019) Probabilistic investigations into the value of information: a comparison of condition-based and time-based maintenance strategies. Ocean Eng 188(July). https://doi.org/10.1016/j.oceaneng.2019.106181
3. Okoh C, Roy R, and Mehnen J (2017) Predictive maintenance modelling for through-life engineering services. Procedia CIRP 59(TESConf 2016), pp 196–201. https://doi.org/10.1016/j.procir.2016.09.033
4. Marvin_Rausand (2004) [Marvin_Rausand,_Arnljot_Høyland]_System_Reliabil(BookFi.org).pdf

5. Pintelon LM, Gelders LF (1992) Maintenance management decision making. Eur J Oper Res 58(3):301–317. https://doi.org/10.1016/0377-2217(92)90062-E
6. Vernier J, M AINTENANCE Méthodes et organisations
7. Kim J, Ahn Y, Yeo H (2016) A comparative study of time-based maintenance and condition-based maintenance for optimal choice of maintenance policy. Struct Infrastruct Eng 12(12):1525–1536. https://doi.org/10.1080/15732479.2016.1149871
8. Cullum J, Binns J, Lonsdale M, Abbassi R, Garaniya V (2017) Risk-Based Maintenance Scheduling with application to naval vessels and ships. Ocean Eng 148(October):476–485. https://doi.org/10.1016/j.oceaneng.2017.11.044
9. Labib AW (2004) A decision analysis model for maintenance policy selection using a CMMS. J Qual Maint Eng 10(3):191–202. https://doi.org/10.1108/13552510410553244
10. Jardine AKS, Lin D, Banjevic D (2006) A review on machinery diagnostics and prognostics implementing condition-based maintenance. Mech Syst Signal Process 20(7):1483–1510. https://doi.org/10.1016/j.ymssp.2005.09.012
11. Banjevic D (2009) Remaining useful life in theory and practice. Metrika 69(2–3):337–349. https://doi.org/10.1007/s00184-008-0220-5
12. Medjaher K, Tobon-Mejia DA, Zerhouni N (2012) Remaining useful life estimation of critical components with application to bearings. IEEE Trans Reliab 61(2):292–302. https://doi.org/10.1109/TR.2012.2194175
13. de Jonge B, Teunter R, Tinga T (2017) The influence of practical factors on the benefits of condition-based maintenance over time-based maintenance. Reliab Eng Syst Saf 158(October):21–30. https://doi.org/10.1016/j.ress.2016.10.002
14. Peng Y, Dong M, Zuo MJ (2010) Current status of machine prognostics in condition-based maintenance: a review. Int J Adv Manuf Technol 50(1–4):297–313. https://doi.org/10.1007/s00170-009-2482-0
15. Si XS, Wang W, Hu CH, Zhou DH (2011) Remaining useful life estimation—a review on the statistical data driven approaches. Eur J Oper Res 213(1):1–14. https://doi.org/10.1016/j.ejor.2010.11.018
16. Bousdekis A, Mentzas G (2017) Condition-based predictive maintenance in the frame of industry 4.0. IFIP Adv Inf Commun Technol 513:399–406. https://doi.org/10.1007/978-3-319-66923-6_47
17. Camci F, Chinnam RB (2010) Health-state estimation and prognostics in machining processes. IEEE Trans Autom Sci Eng 7(3):581–597. https://doi.org/10.1109/TASE.2009.2038170
18. Jiang R (2010) Optimization of alarm threshold and sequential inspection scheme. Reliab Eng Syst Saf 95(3):208–215. https://doi.org/10.1016/j.ress.2009.09.012
19. Silvestri L, Forcina A, Introna V, Santolamazza A, Cesarotti V (2020) Maintenance transformation through Industry 4.0 technologies: a systematic literature review, vol 123. Elsevier B.V.
20. Poor P, Ženíšek D, Basl J (2019) Historical overview of maintenance management strategies: development from breakdown maintenance to predictive maintenance in accordance with four industrial revolutions. In: Proceedings of the international conference on industrial engineering and operations management, no July, pp 495–504
21. C. Krupitzer *et al.*, "A survey on predictive maintenance for industry 4.0," *arXiv*, 2020.
22. Selcuk S (2017) Predictive maintenance, its implementation and latest trends. Proc Inst Mech Eng, Part B: J Eng Manuf 231(9):1670–1679. https://doi.org/10.1177/0954405415601640
23. Poor P, Basl J, Zenisek D (2019) Predictive maintenance 4.0 as next evolution step in industrial maintenance development. Proc-IEEE Int Res Conf Smart Comput Syst Eng, SCSE 2019:245–253. https://doi.org/10.23919/SCSE.2019.8842659
24. Keifer G, Effenberger F (1967) 済無No Title No Title. Angew Chem Int Ed 6(11):951–952
25. Rødseth H, Schjølberg P, Marhaug A (2017) Deep digital maintenance. Adv Manuf 5(4):299–310. https://doi.org/10.1007/s40436-017-0202-9
26. Jasiulewicz-Kaczmarek M, Gola A (2019) Maintenance 4.0 technologies for sustainable manufacturing—an overview. IFAC-PapersOnLine 52(10):91–96. https://doi.org/10.1016/j.ifacol.2019.10.005

27. Sullivan GP, Pugh R, Melendez AP, Hunt WD (2010) Operations & maintenance best practices: a guide to achieving operational efficiency. Fed Energy Manag Program, no August 2010, p 321. http://www1.eere.energy.gov/femp/pdfs/omguide_complete.pdf
28. Zonta T, da Costa CA, da Rosa Righi R, de Lima MJ, da Trindade ES, Li GP (2020) Predictive maintenance in the Industry 4.0: a systematic literature review. Comput Ind Eng 150(August):106889. https://doi.org/10.1016/j.cie.2020.106889
29. You MY, Liu F, Wang W, Meng G (2010) Statistically planned and individually improved predictive maintenance management for continuously monitored degrading systems. IEEE Trans Reliab 59(4):744–753. https://doi.org/10.1109/TR.2010.2085572
30. Nikolic B, Ignjatic J, Suzic N, Stevanov B, Rikalovic A (2017) Predictive manufacturing systems in industry 4.0: Trends, benefits and challenges. In: Annals of DAAAM and proceedings of the international DAAAM symposium, no December, pp 796–802. https://doi.org/10.2507/28th.daaam.proceedings.112
31. Kiangala KS, Wang Z (2018) Initiating predictive maintenance for a conveyor motor in a bottling plant using industry 4.0 concepts. Int J Adv Manuf Technol 97(9–12):3251–3271. https://doi.org/10.1007/s00170-018-2093-8
32. Donovan PO, Gallagher C, Leahy K, Blake S, Bruton K, O'Sullivan DTJ (2017) A systematic mapping of industrial cyber- physical systems research for Industry 4 . 0. In: 34th international manufacturing conference, vol 113, no August. https://doi.org/10.2495/IWAMA150301
33. Yan J, Meng Y, Lu L, Li L (2017) Industrial Big Data in an Industry 4.0 environment: challenges, schemes, and applications for predictive maintenance. IEEE Access 5, no c:23484–23491. https://doi.org/10.1109/ACCESS.2017.2765544
34. Sezer E, Romero D, Guedea F, MacChi M, Emmanouilidis C (2018) An industry 4.0-enabled low cost predictive maintenance approach for SMEs: a use case applied to a CNC Turning Centre," *2018 IEEE International Conference on Engineering, Technology and Innovation (ICE/ITMC)*, pp. 1–8, 2018.
35. Li Z, Wang Y, Wang KS (2017) Intelligent predictive maintenance for fault diagnosis and prognosis in machine centers: Industry 4.0 scenario. Adv Manuf 5(4):377–387. https://doi.org/10.1007/s40436-017-0203-8
36. Sang GM, Xu L, De Vrieze P, Bai Y, Pan F (2020) Predictive Maintenance in Industry 4.0. In: ACM international conference proceeding series. https://doi.org/10.1145/3447568.3448537
37. Lee J, Ni J, Djurdjanovic D, Qiu H, Liao H (2006) Intelligent prognostics tools and e-maintenance. Comput Ind 57(6):476–489. https://doi.org/10.1016/j.compind.2006.02.014
38. Karuppiah K, Sankaranarayanan B, Ali SM (2021) On sustainable predictive maintenance: exploration of key barriers using an integrated approach. Sustain Prod Consum 27:1537–1553. https://doi.org/10.1016/j.spc.2021.03.023
39. Ruiz-Sarmiento JR, Monroy J, Moreno FA, Galindo C, Bonelo JM, Gonzalez-Jimenez J (2020) A predictive model for the maintenance of industrial machinery in the context of industry 4.0. Eng Appl Artif Intell 87(October 2019):103289. https://doi.org/10.1016/j.engappai.2019.103289
40. Guillén AJ, Crespo A, Macchi M, Gómez J (2016) On the role of Prognostics and Health Management in advanced maintenance systems. Prod Plan Control 27(12):991–1004. https://doi.org/10.1080/09537287.2016.1171920
41. Johns T, Street NC (2015) IEEE Standards for prognostics and health management IEEE standards for prognostics and health management. In: 2008 IEEE Autotestcon, no September, pp 97–103
42. Vogl GW, Weiss BA, Donmez MA (2014) Standards for prognostics and health management (PHM) techniques within manufacturing operations. In: PHM 2014—proceedings of the annual conference of the prognostics and health management society, vol 1, pp 576–588
43. Weiss B, Brundage M (2021) Measurement and evaluation for prognostics and health management (PHM) for manufacturing operations—summary of an interactive workshop highlighting PHM trends. Int J Progn Health Manag 12(1):1–19. https://doi.org/10.36001/ijphm.2021.v12i1.2653

44. Shin I et al (2018) A Framework for prognostics and health management applications toward smart manufacturing systems. Int J Precis Eng Manuf-Green Technol 5(4):535–554. https://doi.org/10.1007/s40684-018-0055-0
45. Xia T, Dong Y, Xiao L, Du S, Pan E, Xi L (2018) Recent advances in prognostics and health management for advanced manufacturing paradigms. Reliab Eng Syst Saf 178(June):255–268. https://doi.org/10.1016/j.ress.2018.06.021
46. Zonta T, da Costa CA, da Rosa Righi R, de Lima MJ, da Trindade ES, Li GP (2020) [7]+Failure Mode Identification and Prioritization Using FMECA A Study on Computer Numerical Control Lathe for Predictive Maintenance.pdf. Comput Ind Eng 150(August):106889. https://doi.org/10.1016/j.cie.2020.106889.
47. Atamuradov V, Medjaher K, Dersin P, Lamoureux B, Zerhouni N (2017) Prognostics and health management for maintenance practitioners—review, implementation and tools evaluation. Int J Progn Health Manag 8(Special Issue 7). https://doi.org/10.36001/ijphm.2017.v8i3.2667
48. Aivaliotis P, Georgoulias K, Chryssolouris G (2018) A RUL calculation approach based on physical-based simulation models for predictive maintenance. In: 02017 international conference on engineering, technology and innovation: engineering, technology and innovation management beyond 2020: new challenges, new approaches. In: ICE/ITMC 2017—proceedings, vol 2018-Janua, pp 1243–1246. https://doi.org/10.1109/ICE.2017.8280022
49. Tsui KL, Chen N, Zhou Q, Hai Y, Wang W (2015) Prognostics and health management: a review on data driven approaches. Math Probl Eng 2015. https://doi.org/10.1155/2015/793161
50. Okano MT (2017) IOT and industry 4.0: the industrial new revolution. In: ICMIS-17—international conference on management and information systems, no September, pp 75–82
51. Pivoto DGS, de Almeida LFF, da Rosa Righi R, Rodrigues JJPC, Lugli AB, Alberti AM (2021) Cyber-physical systems architectures for industrial internet of things applications in Industry 4.0: A literature review. J Manuf Syst 58(December):176–192. https://doi.org/10.1016/j.jmsy.2020.11.017
52. Stock T, Seliger G (2016) Opportunities of sustainable manufacturing in industry 4.0. Procedia CIRP 40(Icc):536–541. https://doi.org/10.1016/j.procir.2016.01.129
53. Da Xu L, Xu EL, Li L (2018) Industry 4.0: State of the art and future trends. Int J Prod Res 56(8):2941–2962. https://doi.org/10.1080/00207543.2018.1444806
54. Al-Najjar B, Algabroun H, Jonsson M (2018) Maintenance 4.0 to fulfill the demands of industry 4.0 and factory of the future. J Eng Res Appl 8(11):20–31. https://doi.org/10.9790/9622-0811022031
55. L. an Silvestri, "Maintenance transformation through Industry 4.0 technologies: A systematic literature review," *Procedia CIRP*, vol. 40, no. 11, pp. 3–37, 2020, doi: https://doi.org/10.1016/j.compind.2020.103335.
56. Cachada A et al (2019) Using internet of things technologies for an efficient data collection in maintenance 4.0. In: Proceedings—2019 IEEE international conference on industrial cyber physical systems, ICPS 2019, pp 113–118. https://doi.org/10.1109/ICPHYS.2019.8780217
57. Forcina A, Introna V, Silvestri A (2021) Enabling technology for maintenance in a smart factory: a literature review. Procedia Comput Sci 180(2019):430–435. https://doi.org/10.1016/j.procs.2021.01.259
58. Borangiu T, Morariu O, Răileanu S, Trentesaux D, Leitão P, Barata J (2020) Digital transformation of manufacturing. Industry of the future with cyber-physical production systems. Rom J Inf Sci Technol 23(1):3–37
59. Bag S, Pretorius JHC, Gupta S, Dwivedi YK (2021) Role of institutional pressures and resources in the adoption of big data analytics powered artificial intelligence, sustainable manufacturing practices and circular economy capabilities. Technol Forecast Soc Chang 163(May 2020):120420. https://doi.org/10.1016/j.techfore.2020.120420
60. Durbhaka GK, Selvaraj B (2021) Convergence of artificial intelligence and internet of things in predictive maintenance systems – a review. Turk J Comput Math Educ 12(11):205–214. https://turcomat.org/index.php/turkbilmat/article/view/5862
61. Hartmann D, Van der Auweraer H (2021) Digital twins. In: SEMA SIMAI Springer Series, vol 5, pp 3–17. https://doi.org/10.1007/978-3-030-61844-5_1

62. Tao F, Zhang H, Liu A, Nee AYC (2019) Digital twin in industry: state-of-the-art. IEEE Trans Industr Inf 15(4):2405–2415. https://doi.org/10.1109/TII.2018.2873186
63. Sharma A, Kosasih E, Zhang J, Brintrup A, Calinescu A (2020) Digital twins: state of the art theory and practice, challenges, and open research questions, pp 1–19. http://arxiv.org/abs/2011.02833
64. Assad Neto A, Ribeiro Da Silva E, Deschamps F, Pinheiro De Lima E (2020) Digital twins in manufacturing: an assessment of key features. Procedia CIRP 97:178–183. https://doi.org/10.1016/j.procir.2020.05.222
65. Damjanovic-Behrendt V, Behrendt W (2019) An open source approach to the design and implementation of Digital Twins for Smart Manufacturing. Int J Comput Integr Manuf 32(4–5):366–384. https://doi.org/10.1080/0951192X.2019.1599436
66. Shao G, Jain S, Laroque C, Lee LH, Lendermann P, Rose O (2019) Digital twin for smart manufacturing: the simulation aspect. In: Proceedings—winter simulation conference. vol 2019-Decem, no Bolton, pp 2085–2098. https://doi.org/10.1109/WSC40007.2019.9004659
67. Rosen R, Von Wichert G, Lo G, Bettenhausen KD (2015) About the importance of autonomy and digital twins for the future of manufacturing. IFAC-PapersOnLine 28(3):567–572. https://doi.org/10.1016/j.ifacol.2015.06.141
68. Rasheed A, San O, Kvamsdal T (2019) Digital twin: values, challenges and enablers, 1–31. http://arxiv.org/abs/1910.01719
69. Qi Q, Tao F (2018) Digital twin and big data towards smart manufacturing and industry 4.0: 360 degree comparison. IEEE Access 6:3585–3593. https://doi.org/10.1109/ACCESS.2018.2793265
70. Seliger G, Kim HJ, Kernbaum S, Zettl M (2008) Approaches to sustainable manufacturing. Int J Sustain Manuf 1(1–2):58–77. https://doi.org/10.1504/IJSM.2008.019227
71. Lee JKY et al (2021) Sustainability-oriented application of value stream mapping: a review and classification. IEEE Access 9:68414–68434. https://doi.org/10.1109/ACCESS.2021.3077570
72. Jasiulewicz-Kaczmarek M, Drozyner P (2011) Maintenance management initiatives towards achieving sustainable development. Environ Sci Eng (Subseries: Environmental Science), pp 707–721. https://doi.org/10.1007/978-3-642-19536-5_55
73. Rashid SHA, Evans S, Longhurst P (2008) A comparison of four sustainable manufacturing strategies. Int J Sustain Eng 1(3):214–229. https://doi.org/10.1080/19397030802513836
74. Franciosi C, Voisin A, Miranda S, Iung B (2020) Integration of I4.0 technologies with maintenance processes: what are the effects on sustainable manufacturing? IFAC-PapersOnLine 53(3):1–6. https://doi.org/10.1016/j.ifacol.2020.11.001
75. A. Bastas, "Sustainable manufacturing technologies: A systematic review of latest trends and themes," *Sustainability (Switzerland)*, vol. 13, no. 8, 2021, doi: https://doi.org/10.3390/su13084271.
76. S. E. E. Profile (2019) Sustainability: orientation in maintenance management—theoretical sustainability: orientation in maintenance management—theoretical background, no. August 2013. https://doi.org/10.1007/978-3-642-23553-5
77. Hami N, Shafie SM, Omar S, Ibrahim YM, Abdulameer SS, Muhamad MR (2020) A review of sustainable maintenance in the manufacturing companies. Int J Supply Chain Manag 9(3):935–944
78. Singh RK, Gupta A (2020) Framework for sustainable maintenance system: ISM–fuzzy MICMAC and TOPSIS approach. 290(1–2)
79. Jasiulewicz-Kaczmarek M (2013) Sustainability: orientation in maintenance management: case study, 135–154. https://doi.org/10.1007/978-3-642-23553-5_9
80. Jasiulewicz-Kaczmarek M, Legutko S, Kluk P (2020) Maintenance 4.0 technologies—new opportunities for sustainability driven maintenance. Manag Prod Eng Rev 11(2):74–87. https://doi.org/10.24425/mper.2020.133730
81. T. Mom, "Modernize the shop floor with Manufacturing Operations Management," pp. 1–20.
82. Bakri A et al (2021) Addressing the issues of maintenance management in smes: towards sustainable and lean maintenance approach. Emerg Sci J 5(3):367–379. https://doi.org/10.28991/esj-2021-01283

Smart Inspection; Conceptual Framework, Industrial Scenarios, and Sustainability Perspectives

Ilyasse Houban, Ahmad Aminzadeh, Sasan Sattarpanah Karganroudi, and Noureddine Barka

1 Introduction

Metal manufacturing knew such an evolution these recent years that it made possible the creation of such complex parts to answer the needs of many different industrial sectors among automobile and aerospace which are the most demanding. These parts are posited complex by their geometric details, features, and innovative shapes. In addition, like every manufactured part, they need to go under the quality control process that consists of dimensional and geometrical controls to make sure that the part corresponds to the specifications that were indicated before the manufacturing and to ensure the functionality of the product. Manufacturing companies give significant importance to this process which makes them keep their place in competitive markets by producing high-quality components [1]. However, in the modern era where the industries reduced so much their manufacturing time, the inspection of the parts is the most time-consuming and usually needs human intervention. Even though research on rigid parts inspection reduces the cost and operation time of the inspection by digital tools, the inspection of non-rigid parts remains still an important challenge. For example, a test set of non-solid components in Bombardier Aviation™ as a large industrial company requires 60 to 75 h of operation [2–4]. Therefore, an accurate inspection in a short period became a necessity for manufacturing companies without omitting the cost. And the highlight of this evolution was the appearance of the Computer-Aided Inspection (CAI) allowing for easy 3D scanning of a product to create a digital replica of it stored as a point cloud. The CAI

I. Houban · A. Aminzadeh · N. Barka
Département de Mathématique, Informatique et Génie, Université du Québec à Rimouski, Rimouski, Quebec, Canada G56 3A1

S. Sattarpanah Karganroudi (✉)
Department of Mechanical Engineering, Université du Québec à Trois-Rivières, Drummondville, Quebec, Canada J2C 0R5
e-mail: sattarpa@uqtr.ca

H. Gholami et al. (eds.), *Sustainable Manufacturing in Industry 4.0*,
https://doi.org/10.1007/978-981-19-7218-8_7

opened doors for a new way of inspection, less time- and money-consuming, reducing the human intervention by a large margin. A novel method of real-time 3D scanning of aluminum 5052-H32 laser-welded blanks as semi-flexible parts' inspection based on lean manufacturing concept has been already presented by our research team [5]. Finite element analyses and simulations are also applied in non-rigid inspections to define the influence of process parameters in laser material processing and optimization of objective functions based on genetic algorithms and metaheuristic approaches [6–8]. The CAI uses advanced data acquisition methods instead of traditional and tactile tools such as calipers, optical comparators, and gauges. The CAI methods are capable of creating a point cloud of inspection surfaces using contact or contactless scanners to fulfill the inspection that is most of the time non-destructive. Back in time, the quality inspection was usually done by using micrometers, calipers, gauges, optical comparators, and manual Coordinate Measuring Machines (CMM) but those processes are slow and require a lot of programming and planning before beginning to function. They also need contact with the fixture and can be a redundant and tedious process to double-check practical errors. Over the past few years, digital data acquisition devices have been advancing drastically and rapidly such as 3D optic and laser scanners [9, 10] along with computational algorithm and calculation developments that made possible the use of computer-aided inspection (CAI) methods. In this regard, 3D data acquisition devices make it possible to obtain the coordination of points on the inspection surfaces of parts, called point clouds, by scanning the surface of parts during the inspection process. From point clouds obtained a scan mesh is generated, simplified, and optimized using mesh smoothing and decimation methods [11, 12]. The objective of the scan mesh is to represent the geometrical shape of the part in the most accurate way with the least required data volume which can be translated to the mesh size. CAI methods help to make an automatic time-saving inspection by both applying tolerancing methods and computational meshing tools. Those methods make it easy to compare a computer-assisted model (CAD) with a photocopying mechanism in a standard communication system to define the geometric deviations from the surface of the parts during the production process. The issue related to this test is the link system. In fact, the CAD model is located in the Design Coordinate System (DCS) and the scanning data is in the Measurement Coordinate System (MCS), and the registration systems built into the computer-assisted registration are required to synchronize the two types in one system.

Depending on the type of test components such as simple or complex geometry, rigid or non-rigid behavior, industrial applications, and tolerance requirements, a variety of computer-assisted methods can be found. On one hand, solid registrations are used for solid components to adapt the scanning model in a free environment in relation to the CAD model. On the other hand, non-solid parts know complex registration as the deviation of the geometric parts can exceed tolerance due to the compatibility of the components in the free state. This complication opens the door to two categories of non-rigid part inspection: physical fixtures inspection methods and fixtureless non-rigid registration methods. Operational limits and drawbacks of using fixtures encouraged industrial sectors towards using fixtureless inspection methods that firstly appeared in 2002 [13]. Figure 1 shows a framework hierarchy for automated inspection using a 3D scanner [14]. Here, rigid and non-rigid parts have been discussed in detail and then a novel digital model is presented for parts' inspection in industrial application.

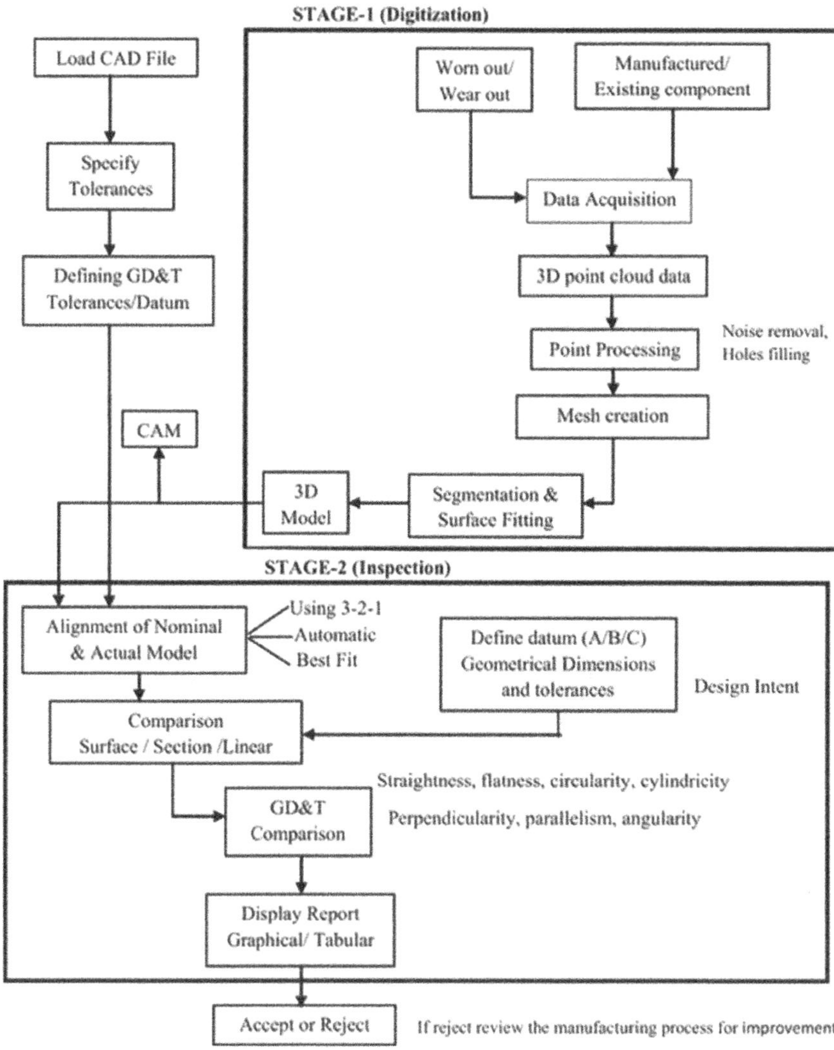

Fig. 1 Framework hierarchy for automated inspection using a 3D scanner

2 Data Acquisition Methods

The technical drawing of each manufactured part goes through a lot of investigation and steps before its validation and acceptance from the engineers to make sure that any misunderstanding of the parameters has been cleared. It is the first step in the manufacturing process. This is done to make sure that the final product will respond to the different dimensional and geometrical criteria defined such as the dimensions, angles, and precision sought for each part of the workpiece. Even though all those actions are taken to make sure that the manufacturing of the part will be as precise as we can, we can't omit the importance of the measurement and data acquisitions of the workpiece to make it possible to validate the manufacturing work. Thus, the biggest issue the modern manufacturing field is facing is that the traditional measurement and data acquisition methods require human intervention, skillful operators, and most of all they are time-consuming. Some methods were developed to increase the efficiency of the 3D topography of surfaces like the spiral sampling [15] but with the increasing demands of customers that not only expect higher quality, lower price, and higher performance, but also require the earliest delivery of the product, better and less time-consuming methods had to be chosen.

Advances in 3D scanning technology allow the creation of a digital scanning model from a visual object. Advanced measurement systems and specific scanners can be classified as contact and non-contact scanners. Contact scanners knew their origins in 1933 by Abbot and Firestone [16]. The affected profilometer is still the most common sharpness measuring device in the machinery industry and has been developed over time. Those scanners are now based on Coordinate Measuring Machine (CMM) technology that can be controlled manually or automatically by the system. These devices contain a three-axis probe, in which each axis has a built-in reference (Fig. 2a). Communication scanners are very reliable as they are not sensitive to color or visibility and are more accurate and less expensive than certain communication scanners. However, data acquisition from these devices is slow and the probe contact may affect certain non-solid components that cause unwanted flexibility during the scanning process [17]. Non-contact scanners use lasers and optics (e.g., using integrated device sensors (CCD) introduced in Fig. 2b) to scan numerically the geometrical position of the part as point clouds. Unlike contact scanners, those devices are faster while there is no physical contact between the scanner and the operating environment. The accuracy of the data obtained by non-contact scanners is low compared to contact scanners but still, the accuracy is acceptable for industrial testing systems. As flexible as they may seem to like, non-contact scanners have some limitations and disadvantages. Highlighting, display, or location color may affect the data acquisition of these devices. The limitations of non-contact scanners can be reduced by using temporary paint that does not show to make the scan more reliable [18]. These limitations can add noise to the cloud of available points where the validity of appropriate testing methods needs to be validated.

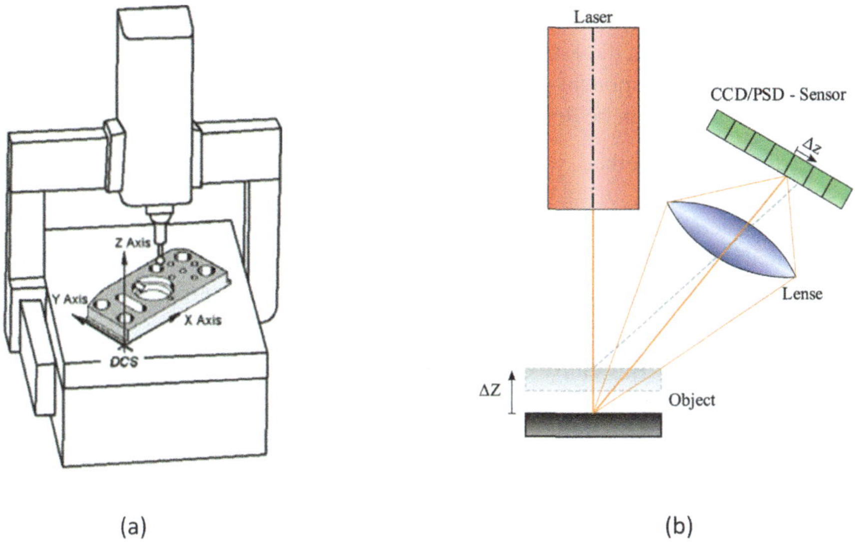

Fig. 2 **a** Contact scanner based on CMM [19] **b** principle of a laser triangulation sensor [20]

3 Quality Control Based on 3D Geometrical Inspection

Almost every industry, at different levels, is required to use quality control. No matter what kind of industry or job, each company needs to evaluate its production quality to improve its product, continue to compete, and maintain a good reputation. Quality control not only removes incomplete components but also ensures a complete evaluation of the product quality of the product. Assessment tasks can be divided into the following three broad categories [21] where general tasks are listed in each category:

- Gross inspection: This is the first step of inspection as it is done with the bare eye to obtain diagnostic information of the workpiece. It's done by comparing the visual form of a workpiece to its CAD model [22].
- Dimensional and geometrical inspection: In this step, different dimensions of a workpiece are measured to verify if they respond to the tolerance requirements [23].
- Micro inspection: This step inspects the manufacturing quality of the workpiece to verify the integrity of the parts such as the surface roughness and porosity [24].

Quality control is an important function in the lifecycle of mechanical products, especially in a seamless manufacturing process. To date, components and organizations must be tested to ensure that they meet their definitions. The test results provide important information about the behavior of production processes. For example, an intolerance hole could indicate that the cutter had removed the worm and should be replaced. Tolerancing is the process of ensuring partial exchange by controlling the geometrical dimensioning variations present in the fabricated parts. Tolerance comes from specifying the extent to which the size is allowed to vary. In other words, tolerance ensures component performance and production quality. According to the literature, tolerance can be defined as the following [25]:

- Direct tolerance method, which includes size limitation and addition/subtraction tolerance.
- Typical tolerance notes, dealing with the tolerance of all sizes.
- Geometric Dimensioning and Tolerancing (GD&T), to ensure the compatibility of the components made with the sign described in the design phase.

Geometric Dimensioning and Tolerancing (GD&T) is a core aspect of inspection and control quality. Basically, GD&T is a system credited to Stanley Parker who developed the concept of "true position." The purpose of geometric dimensioning and tolerancing is to define and communicate engineering tolerances. It makes it possible to define the allowable variation by describing the nominal geometry of CAD and engineering drawings using symbolic language. In the manufacturing and production industry, GD&T is widely applied in the manufacturing field for workpieces with complex shapes in different industrial disciplines. Generally, all types of industrial parts could be categorized as rigid and non-rigid parts. This division opens the door to a variety of definitions and approaches to tolerance. Methods for tolerating non-solid components should be considered, compliant, and permitted to deliver such material during the evaluation and performance of the organization. It first appeared in 1996 [26] in the construction of high-end vehicles, and the tolerance analysis of non-solid parts appeared in time. In this context, the tolerance of the profile is given a place in the free form of components to control land diversity. This profile tolerance can be defined in terms of the datum (names) known as related profile tolerances. Related profile tolerances are used in cases involving a combination of free-form spaces and other geometric features [27]. Once tolerance is allocated, geometric and size requirements must be confirmed in the testing process component. Figure 3 shows the classification of certain specification methods used for geometric sizes and tolerance for non-solid components [28].

In the era of automation, production standards such as ASME Y14.5 and ISO-GPS were developed to improve the quality and standard of GD&T. Also, the American Society of Mechanical Engineers (ASME) has developed rules, definitions, requirements, defaults, and recommended practices to make that happen. They say that equipment and parts should be tested in a free form that is represented by a mark on the drawings. However, testing of non-solid parts should take into account the compliance with the defects of these components. As a result, specific requirements, based on ASME Y14.5 and ISO-GPS standards, for geometric reduction and tolerance for non-solid components have been developed as shown in Fig. 4.

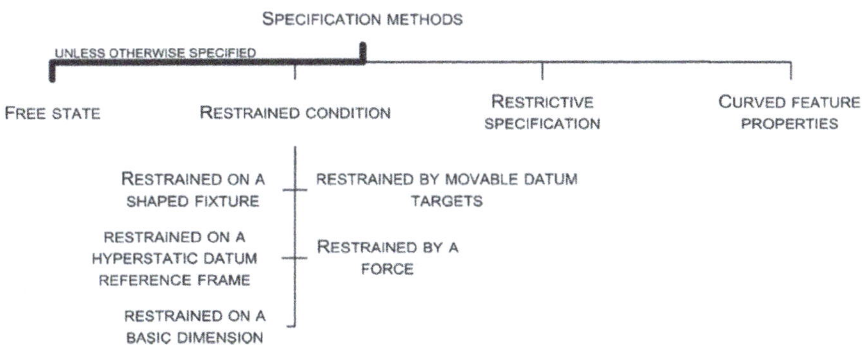

Fig. 3 Categorization of particular specification methods used for the geometric dimensioning and tolerancing of non-rigid parts

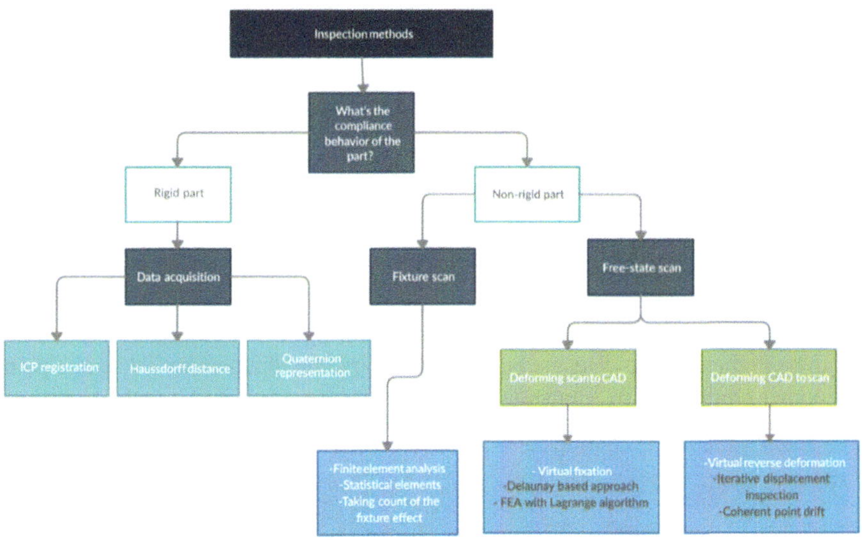

Fig. 4 Graphical representation of existing inspection methods

4 CAD Model and Scan Data Registrations

The main objective of computer-aided inspection (CAI) is to make it possible to compare the computer-aided design (CAD) with the data acquired from methods cited in the previous section. Workpieces being categorized as rigid and non-rigid, different registration methods have been developed to make it possible to inspect every part. In fact, rigid registration is the primary step in computer-aided inspection for non-rigid parts. The rigid and non-rigid registration methods are discussed in detail in the following sections.

4.1 Rigid Registration

The main purpose of the robust registration is to bring CAD and scanning models as close as possible to the standard communication system without disabling both types [2]. It uses the conversion matrix to effectively translate and rotate models without making changes to their structure. Historically, CAI first appeared in 1992 by Besl and McKay [29], the Iterative Close Point (ICP) algorithm is one of the strongest and most robust registrations. Although different approaches emerged over the years such as those presented by Li and Gu in 2004 [19] or Savio in 2007 [30], ICP is still widely applied in different domains, for example for an inspection of an aircraft [31], it has its place among the most reliable and statistically robust methods of registration. Figure 5 shows the result of an ICP algorithm when applied on CAD and scan models of a workpiece that are not in the same coordinate.

The ICP algorithm can be applied using the four following steps:

- Match each point from the point cloud set to the closest point in the reference set (CAD).
- Estimate the combination of rotation and translation to find the transformation matrix.
- Transform the source points using the obtained matrix.
- Iterate (re-associate the points).

A transformation matrix that includes translation and rotation in ICP registration is measured and calculated in each iteration to reduce the distance between the two types. The key calculation used in this algorithm is the Hausdorff distance [32]. It measures the distance between CAD mesh and cloud point data obtained for scanning. It can be defined as the distance between all the points of an empty set to one point of another empty set. We can show that in Eq. 1 where dH (X, Y) is the Hausdorff range and (X, Y) are two empty subsets.

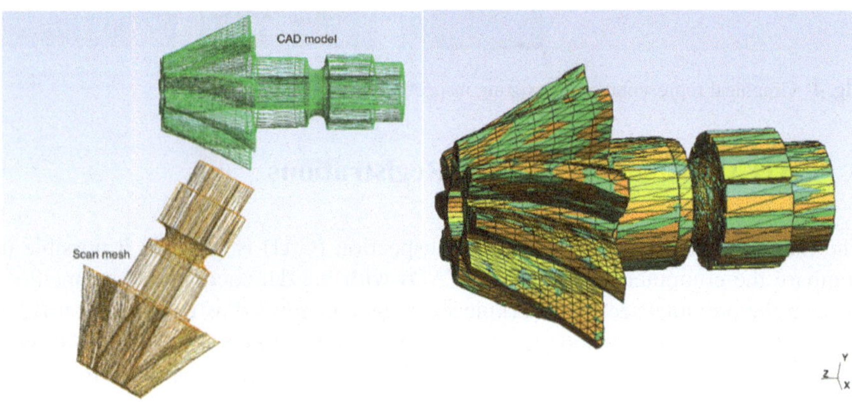

Fig. 5 Result of ICP applied on a CAD (green) and its scan (orange) model

$$d_H(X, Y) = \max\left\{\sup_{x\in X}\inf_{y\in Y} d(x, y), \sup_{y\in Y}\inf_{x\in X} d(x, y)\right\} \tag{1}$$

An algorithm, as strong as ICP and as used as it, has definitely been subject to a lot of improvements and upgrades. The evolution of the different fields where ICP is being used had made it clear that we can't use the same algorithm that appeared nearly three decades from now. The ICP has been modified and improved to reduce the initial calculation time in 1995 [33] by proposing a more robust approach to using random cloud point models. The algorithm also knew the strategy of reducing [34] or reducing the error of the transformation metrics. Advances in the search for the closest points using the corresponding points from previous ICP duplication and search only in a small area near those points have made a significant improvement in the processing time of the algorithm [35]. The registration process also has been improved by some techniques to accelerate the process and upgrade efficiency [36]. Color registration has been also implemented to improve the efficiency of the transformation of the sets [37], although not all the scanners can acquire the colors from the workpieces, this variant of the algorithm can't be ignored. The ICP knew many more improvements, and many variants for the algorithm have been investigated [34].

4.2 Non-rigid Registration

Rigid registration, which is the first step in the registration process, is not sufficient and is not a reliable method to be used on non-solid applications as they have many limitations to be kept in mind. Considering the flexibility of the components in the free state, comparisons between CAD and scanning models cannot identify defects and measure their size in the scanning model. To solve this problem, CAI methods for non-solid components are used to distinguish between errors, such as geometric deviations and distortions in relation to the CAD model, as well as flexibility due to compliance with non-solid components. As already mentioned, the standard methods of mitigation and testing of non-solid components set up over-restricted testing materials to compensate for the dynamic variability of these components and to ensure that the standard setting best reflects the performance of the session.

Before introducing non-solid registries and their methods, a better understanding of the compatibility of non-solid components is required. The conceptual definition (coherence) of non-solid parts is related to the flexibility and geometry of the components. In fact, the higher the degree of coherence of the components means the higher flexibility of these components. Therefore, the flexibility of the non-solid components in the free state is due to the consistent behavior of these components. By looking at the information of the completed analysis, [K] {u} = {f}, corresponding (C) is defined in Eq. 2

$$C = \{u\}^t\{f\} \tag{2}$$

where $\{f\}$ is the force vector, $[K]$ is the global stiffness matrix, and $\{u\}$ is the displacement vector. The flexibility is defined as the inverse of stiffness ($[K]^{-1}$) accordingly.

Due to the behavior of the non-rigid parts, it is clear that their compliance will affect the inspection process. The deformations caused by their weights will get in the way of the registration of the workpieces. To counter this behavior, fixtures and jigs are used to put those non-rigid parts in their assembly position. This way the inspection process will be simpler, and we can register the workpiece the same way we do with a rigid part. However, a number of downsides exist in using fixtures such as their time-consuming setup process, considerable acquisition and operation expenses, and limitations of standard fixtures in some scenarios. The companies find themselves in the obligation to design and manufacture costly conformation jigs to try to recreate as much as possible the assembly state of the parts. That being said, those disadvantages made it obvious that fixture registration was not the best solution. Researchers have tried to avoid the use of those fixtures by numerically deforming the data acquired by the scan until it matches the CAD or vice versa. Thereby elastically deforming the data to reach an optimal assembly shape while avoiding any manufacturing defects of the jigs. The fixtureless CAI methods are divided into four modes as (i) automatic vision inspection, (ii) metric factor, (iii) boundary reconstruction, and (iv) simulated displacement. Fixtureless testing of non-solid parts can be done by non-solid registration methods classified as simulated displacement. These methods are actually based on compensating for the flexibility of non-solid components in a free environment with optical migration. The main idea of the non-invasive methods is to enable comparisons between scan and CAD models by compensating for the variable variability of the part while leaving the error areas firm.

5 Intelligent Factory Based on Computer-Aided Inspection (CAI)

In the context of Industry 4.0, inspection is a fundamental stage towards sustainable manufacturing in industrial applications. In fact, the importance of this process has already been mentioned above as it turns physical parts into information to make it possible for manufacturers to evaluate the quality of the parts made and their conformity to specifications pre-defined on the CAD model. Inspection is among the main contributors to the value of a product. The main concern of Industry 4.0 being time management and cost reduction, planning every task is then critical to assure the best results. As a matter of fact, inspection planning is popping out as a key element for the upgrade of Inspection 4.0. Many advanced methodologies for sampling strategy design have been thought of as the current practices, in general, see the operator as the main actor, and this is due to the lack of information circling both at a system level and for the machine during the measurement. Intelligence

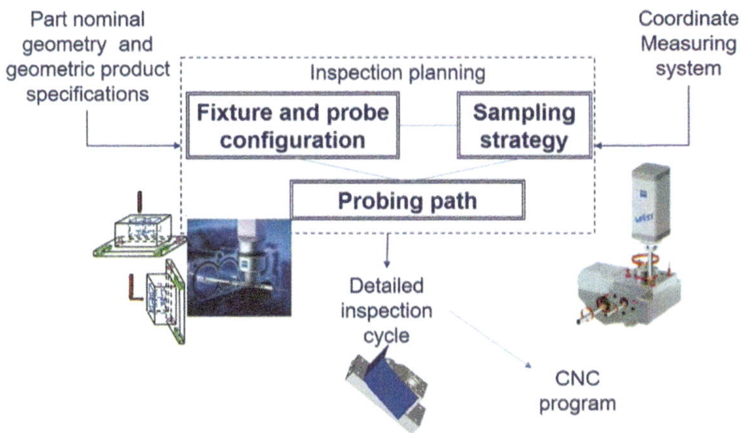

Fig. 6 Inspection planning in coordinate measurement system

and mostly artificial intelligence (AI) is needed to optimize the sampling strategies. From those methodologies, we can name the Point Distribution, Choice of sample size, and Path planning and probe configuration [38]. Figure 6 shows, as an example, inspection planning in the coordinate measurement system.

Reducing the human intervention in the era of Industry 4.0 gave way to the implementation of industrial robotic arms with server computers, sensors, and actuators in a way it can be useful in any field, making it possible for the automation of non-destructive testing (NDT) [39]. The process of data acquisition can be very redundant and human intervention can make mistakes and is for sure not as precise and accurate as a robotic arm controlled by a computer. A part can be sent to the NDT facility with some basic information about the piece, the data will be collected automatically and the best way of scanning/inspecting (type of probe, pathway, registration method) will be exported from the database, and the robotic arms and computers will apply the pre-defined sequences of actions to inspect the part. Basically, these elements are the foundation of an intelligent factory in the concept of Industry 4.0.

5.1 *Digital Twins (DT)*

First and foremost, scanning inspection is among the main contributors to value not only for a product but also for a whole factory. As being said, it is definitely one of the key processes to upgrade during the era of Industry 4.0 as this revolution's goal is to automate the traditional manufacturing and industrial practices using modern smart technologies. Human intervention is being less and less required as it was mentioned above for its time-consuming tasks. Recently, a numerical solution has been presented on product lifecycle management at the University of Michigan Lurie Engineering Center as Digital twins (DT) [40]. There is a correlation relation between CAI and

DT model. In this regard, Airbus A350™ is an example of advanced aircraft which is entirely based on a 3D digital mock-up. This technology is facilitating a significant decline in progress time and development. Generally, the first stage is to generate a virtual version of the asset that is known as a digital twin. Moreover, all along its cycle, with intelligent and dynamic data modeling. In Fig. 7 is illustrated a basic representation of how DT can help manufacturing factories. A twin model manages data in an extra robust approach for operators to easily identify essential simulations, reports, device history, and results. Also, by providing a virtual interpretation of the product lifecycle, it permits us to make better decisions and predict problems. Basically, the digital twin concept is based on three steps, 3D definition, 3D in context, and 3D as a service. The first step includes using the best technology and process to obtain the real world and create a 3D definition by generating point clouds via 3D laser scanning. Then, the 3D model should be capable of evolutions, a variety of revisions, and structures of the asset or product. Finally, the model should be customized with 3D functionalities to recommend services to each actor of the product lifecycle in a way that will improve operational outcomes.

The digital twin can be utilized for any asset, equipment, or machine in a factory or complete factory [41]. The real idea of Digital Twin began from product lifecycle management (PLM) earlier than new technologies such as the Internet of Things, Smart Manufacturing, and Industry 4.0. Twin's structure has multiple elements to simulate states and then future circumstances of the manufacturing process. In current trends, sensors play a key role to collect any sort of data in real-time conditions. The smart network software has been developed and tried to visualize physical plants in the digital world. Incidentally, networks are diverse from various wired and wireless networks. Now, the digital twin has major questions related to a user persona, value chain, its supporting data model, inputs, user interfaces, benefits, and supporting business models. It is worth mentioning that the current factory system simulation is

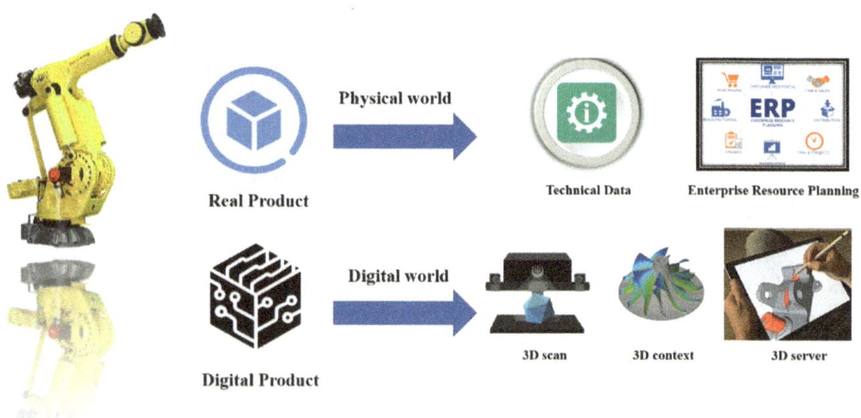

Fig. 7 A general model of digital twin

based on four stages namely machine state monitoring, forklift movements, constraint management, asset position, and orientation. All these data should be transmitted by a real-time network connection which provides whole information for different types of the value chain in manufacturing sectors. In this regard, in our model, Internet of things is considered as the core of data transmission from all the units.

5.2 *The Internet of Things (IoT)*

The industry knew such a development lately in all its different aspects and with the upcoming 4.0 revolution being one big step towards the future many technologies had to keep up with its pace. One of the key elements of Industry 4.0 development is the Internet of things. It may be described as the network of physical objects. The Internet itself was one of the greatest achievements of humankind as it made it possible to connect people all over the globe and made access to information so much faster and easier. Due to its importance, it seems that anything related to the Internet may just be as beneficial to the development of modern technologies. The Internet of things (IoT) serves the same purpose as the "common" Internet as it connects objects that are embedded with software, sensors, machine learning, real-time analytics, and multiple other technologies for the purpose of exchanging data with other devices and systems over the cloud. The cloud being a non-physical network to gather the information and made it for storage, it is also easy to access, and the follow-up can be done in a matter of seconds. The most common example of IoT is smart houses; it's an environment that you can control by a single device. A smartphone can turn off the lights, open the windows, or heat up the oven with clicks. It seemed futuristic at first, but it became reality in no time. Now imagine all the machines of a factory controlled from a distance with a single device and with machine learning those machines can also be automated and gain autonomy at some point. With this kind of technology, the privacy and security of the industry can also be upgraded. In the industry, all the modern machines have sensors that acquire so much information in a very short time but mostly those machines are not related one to another and it takes time to gather all the information of a factory. Apart from that, some tasks are redundant, take a lot of time, and human error can affect the results a lot because the appearance of robots in different steps of the industry became so much important to its development. IoT can make it one step further as it can make the usage of machines and robots so much more efficient and precise. New opportunities and possibilities will surely be available thanks to IoT as it is a new field that knows no barriers or limitations due to the aspect of the Internet being so vague and limitless.

An example of an intelligent factory will be presented to give an idea of how things might evolve later. We will separate a product life into four different steps and explain how the IoT can revolutionize that. The inventory, the manufacturing, the inspection, and then the customer. Thanks to IoT a cloud network will connect all those different steps assuring constant feedback, information storage, and communication. Firstly, the design of the product will be made responding to the demands of the customer

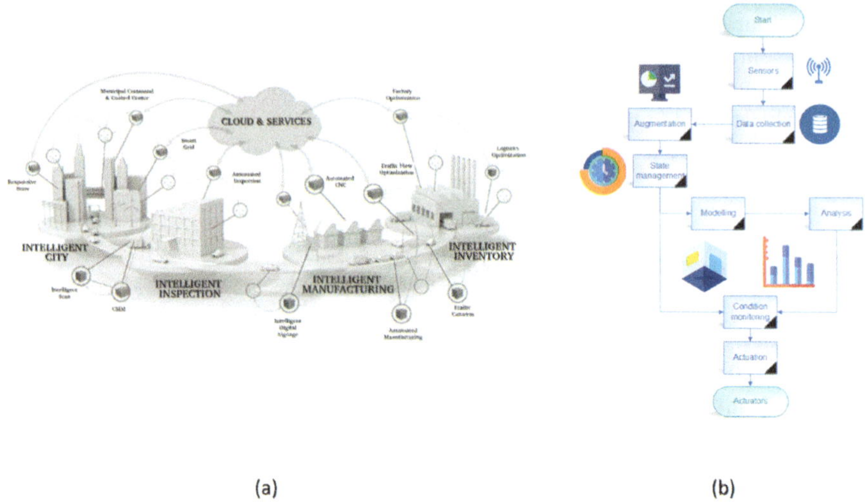

Fig. 8 **a** Digital factory based on industry 4.0 concept. **b** Digital twin data flow from device

and then sent to the network. A small check in the inventory will be the next step to see the availability of the raw materials that will be sent to the manufacturing process. The design that had already been analyzed made it possible to extract a manufacturing protocol that will be sent to the machines to create the product. The next step will be the inspection, and like in the manufacturing process, the data collected will dictate which kind of inspection methods and tools will be used to acquire the best scan possible. The material, rigidity, and size of the product are the most important factors in this step. After the inspection, the data is analyzed, and if everything checks, the product is sent to the customer. All these steps are automated without any need for human intervention. The time-consuming and repetitive aspect of some actions will be reduced this way and the human error will be completely erased. At this point, the only human work that is left might be the maintenance of the different machines but even that can be at some time delegated to robots as sensors will send the information to the cloud of a failing part in a machine. Figure 8 is a representation of the basics of Industry 4.0 in sustainable manufacturing [42].

5.3 *Sustainable Inspection*

Managing jobs in an environmentally and socially responsible manner—"sustainable manufacturing"—is no longer just a good thing to have, but it is important for business. Companies around the world are facing rising costs in building materials, capacity, and compliance in line with the high expectations of customers, investors, and local communities. Many businesses have begun to take important steps to grow greenery to ensure that their development is economically and environmentally sustainable. Their pioneer experience shows that environmental development goes hand in hand with increased profitability and competitiveness. However, many small and medium manufacturing enterprises have not yet had these wonderful opportunities. They may be struggling to make ends meet, or they may be feeling overwhelmed by the demands of their customers, or they may not have the knowledge and resources to invest in the environment, or they may simply not know where to start.

"Sustainable manufacturing" is the official name for an exciting new way to do business and build value. It is behind many of the green materials and processes. Businesses of all kinds are already involved in new initiatives and initiatives that help promote a healthy environment, increase their competitiveness, reduce risk, build trust, drive money, attract customers, and make a profit. There is no single definition for sustainable manufacturing but the United States Department of Commerce's Sustainable Manufacturing Initiative summarizes it this way: "The creation of manufactured products that use processes that minimize negative environmental impacts, conserve energy and natural resources, are safe for employees, communities, and consumers and are economically sound." Simply put, sustainable manufacturing is all about reducing the various business risks that come with any manufacturing activity while maximizing new opportunities arising from improving your processes and products.

There is no clear work done towards the application of sustainability in the inspection field but by applying the aspects presented in Fig. 9, we can present a vision for sustainable inspection [43]. Being a key process to define the quality and efficiency of a product, the inspection should definitely lean towards sustainability in the upcoming years. With the help of the Industry 4.0 revolution, the human intervention will be minimized and that can ensure safety for more employees by not including them in processes that use radiations like the laser inspection or might be redundant and don't provide a great work environment, and in an environmental level, we will be able to minimize the waste and destruction of parts as the manufacturing process will evolve with the predictions for upcoming parts with the help of the smart inspection. The constant feedback between both inspection and manufacturing processes will assure more sustainable results. The possibilities that can be created from this relationship will be a key for the constant improvement of the inspection process towards sustainability. In fact, the improvement in inspection steps can result in lowering the energy and resources consumed during this process and at the same time be money-saving which opens doors to new investments and improvements in the different steps of the lifecycle of a part. It is obvious that sustainable manufacturing and inspection will improve drastically all the aspects of businesses and not just the environmental ones. A combination of the Industry 4.0 revolution and sustainable inspection appears to be the future of the manufacturing industry.

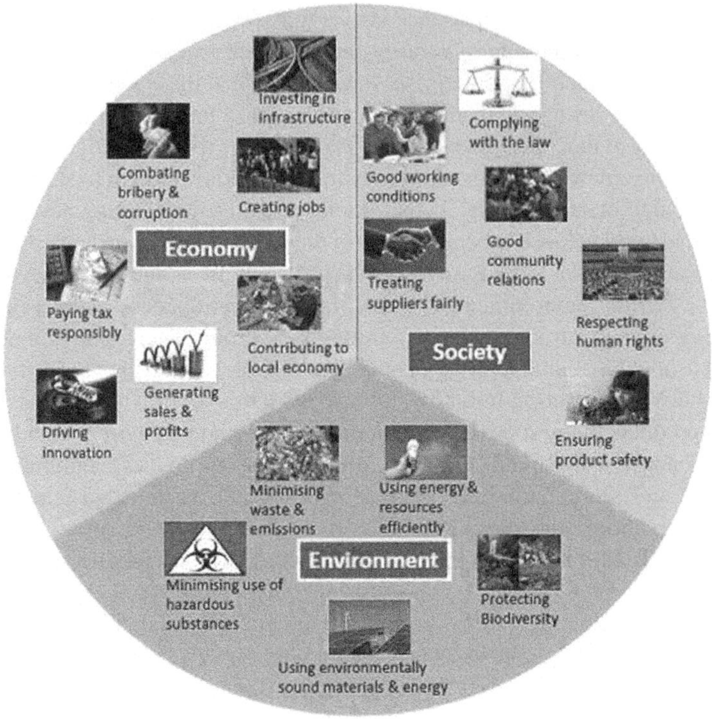

Fig. 9 Three-dimensional aspects of sustainable manufacturing

6 Conclusion

This study proposed a systematic review of the works done in the field of 3D geometric inspection and its importance in the manufacturing industries. The methods cited in this review englobe aspects of inspection in the era of Industry 4.0 with the criteria towards different fields and types of parts. Inspection, as a core of quality control, is one of the most important in the production cycle of a part. In this regard, its development is not to be neglected. As we move more towards Industry 4.0, the implementation of its ideas should be part of the development of dimensional and 3D geometrical inspections that already exist. To this end, the exact data of a manufactured part is acquired and compared to its CAD model to verify whether it meets the assembling and functional requirements without any human intervention. A fully automated and intelligent Inspection 4.0 in its industrial use is in its way wherein an automated factory with the least human intervention is the main goal. Finally, a digital twin's model has been proposed for an industrial application based on Industry 4.0 answering the question that can it be possible at some point to use artificial intelligence to delegate redundant inspection works to machines expecting feedback from them as well as an auto-maintenance at some point?

References

1. Pinjala SK, Pintelon L, Vereecke A (2006) An empirical investigation on the relationship between business and maintenance strategies. Int J Prod Econ 104:214–229
2. Karganroudi SS, Cuillière J-C, Francois V, Tahan S-A (2016) Automatic fixtureless inspection of non-rigid parts based on filtering registration points. Int J Adv Manuf Technol 87:687–712
3. Sattarpanah Karganroudi S, Cuillière J-C, François V, Tahan S-A (2018) "What-if" scenarios towards virtual assembly-state mounting for non-rigid parts inspection using permissible loads. Int J Adv Manuf Technol 97:353–373
4. Karganroudi SS, Cuillière J-C, François V, Tahan S-A (2017) Assessment of the robustness of a fixtureless inspection method for nonrigid parts based on a verification and validation approach. J Verif Valid Uncertain Quantif 2
5. Aminzadeh A, Karganroudi SS, Barka N, El Ouafi A (2021) A real-time 3D scanning of aluminum 5052–H32 laser welded blanks; geometrical and welding characterization. Mater Lett 296:129883
6. Aminzadeh A, Aberoumand M, Rahmatabadi D, Moradi M (2021) Metaheuristic approaches for modeling and optimization of FDM process. In: Dave HK, Davim JP (eds) Fused depos. model. based 3D print, Springer International Publishing. https://doi.org/10.1007/978-3-030-68024-4
7. Aminzadeh A, Parvizi A. Deep drawing of tailor laser welded blanks: experimental and finite element investigation n.d.:1–38
8. Aminzadeh A, Karganroudi SS, Barka N (2021) A novel approach of residual stress prediction in ST-14/ST-44 laser welded blanks; mechanical characterization and experimental validation. Mater Lett 285:129193
9. Zhang S (2010) Recent progresses on real-time 3D shape measurement using digital fringe projection techniques. Opt Lasers Eng 48:149–158
10. Bi ZM, Wang L (2010) Advances in 3D data acquisition and processing for industrial applications. Robot Comput Integr Manuf 26:403–413
11. Karbacher S, Haeusler G (1998) New approach for the modeling and smoothing of scattered 3D data. Three-Dimensional Image Capture Appl 3313:168–177
12. Jones TR, Durand F, Desbrun M (2003) Non-iterative, feature-preserving mesh smoothing. In: ACM SIGGRAPH 2003 Pap., pp 943–9
13. Blaedel K, Swift D, Claudet A, Kasper E, Patterson S (2002) Metrology of non-rigid objects
14. Pathak VK, Singh AK, Sivadasan M, Singh NK (2018) Framework for automated GD&T inspection using 3D scanner. J Inst Eng Ser C 99:197–205. https://doi.org/10.1007/s40032-016-0337-7
15. Wieczorowski M (2001) Spiral sampling as a fast way of data acquisition in surface topography. Int J Mach Tools Manuf 41:2017–2022
16. Abbott EJ (1933) Specifying surface quality. Mech Eng 55:569–572
17. Leake JM, Borgerson JL (2013) Engineering design graphics: sketching, modeling, and visualization. Wiley
18. Lehmann KM, Azar MS, Kämmerer PW, Wentaschek S, Hell ENF, Scheller H (2011) The effect of optical conditioning of preparations with scan spray on preparation form. Acta Stomatol Croat 45
19. Li Y, Gu P (2004) Free-form surface inspection techniques state of the art review. Comput Des 36:1395–1417
20. Park J-S, Jang W-J, Lee S-B, Park C-S (2008) Acquisition model for 3D shape measurement data. Int J Precis Eng Manuf 9:16–21
21. Spyridi AJ, Requicha AAG (1990) Accessibility analysis for the automatic inspection of mechanical parts by coordinate measuring machines. In: Proceedings, IEEE Int. Conf. Robot. Autom., pp 1284–1289
22. Ishiwatari H, Sato J, Fujie S, Sasaki K, Kaneko J, Satoh T et al (2019) Gross visual inspection by endosonographers during endoscopic ultrasound-guided fine needle aspiration. Pancreatology 19:191–195

23. Nguyen CHP, Choi Y (2018) Comparison of point cloud data and 3D CAD data for on-site dimensional inspection of industrial plant piping systems. Autom Constr 91:44–52
24. Schebesch K-M, Doenitz C, Haj A, Höhne J, Schmidt NO (2020) Application of the endoscopic micro-inspection tool QEVO®in the surgical treatment of anterior circulation aneurysms-a technical note and case series. Front Surg 7:101
25. Talebi S, Koskela L, Tzortzopoulos P, Kagioglou M, Krulikowski A (2020) Deploying geometric dimensioning and tolerancing in construction. Buildings 10:62
26. Liu SC, Hu SJ, Woo TC (1996) Tolerance analysis for sheet metal assemblies
27. Li Y, Gu P (2005) Sculptured surface tolerance verification with design datums. Int J Prod Res 43:1465–82
28. Abenhaim GN, Desrochers A, Tahan A (2012) Nonrigid parts' specification and inspection methods: notions, challenges, and recent advancements. Int J Adv Manuf Technol 63:741–752
29. Besl PJ, McKay ND (1992) Method for registration of 3-D shapes. In: Sens. fusion IV Control Paradig. data Struct., vol 1611, pp 586–606
30. Savio E, De Chiffre L, Schmitt R (2007) Metrology of freeform shaped parts. CIRP Ann 56:810–835
31. Ravishankar S, Dutt HNV, Gurumoorthy B (2010) Automated inspection of aircraft parts using a modified ICP algorithm. Int J Adv Manuf Technol 46:227–236
32. Henrikson J (1999) Completeness and total boundedness of the Hausdorff metric. MIT Undergrad J Math 1:69–80
33. Masuda T, Yokoya N (1995) A robust method for registration and segmentation of multiple range images. Comput Vis Image Underst 61:295–307
34. Rusinkiewicz S, Levoy M (2001) Efficient variants of the ICP algorithm. In: Proc. third int. conf. 3-D digit. imaging model, pp 145–52
35. Greenspan M, Godin G (2001) A nearest neighbor method for efficient ICP. In: Proc. Third Int. Conf. 3-D Digit. Imaging Model., p 161–8
36. Zhu L, Barhak J, Srivatsan V, Katz R (2007) Efficient registration for precision inspection of free-form surfaces. Int J Adv Manuf Technol 32:505–515
37. Li X, Du S, Li G, Li H (2020) Integrate point-cloud segmentation with 3D lidar scan-matching for mobile robot localization and mapping. Sensors 20:237
38. Moroni G, Petrò S (2018) Geometric inspection planning as a key element in industry 4.0. In: Int. Conf. Ind. 4.0 Model Adv. Manuf., pp 293–310
39. Mineo C, Vasilev M, Cowan B, MacLeod CN, Pierce SG, Wong C, et al (2020) Enabling robotic adaptive behaviour capabilities for new industry 4.0 automated quality inspection paradigms. Insight-Non-Destructive Test Cond Monit 62:338–44
40. Malik AA, Bilberg A (2018) Digital twins of human robot collaboration in a production setting. Procedia Manuf 17:278–285
41. Lu Y, Liu C, Kevin I, Wang K, Huang H, Xu X (2020) Digital Twin-driven smart manufacturing: connotation, reference model, applications and research issues. Robot Comput Integr Manuf 61:101837
42. Dobrin C (2014) Industry 4.0, M2m, Iot\&S-All Equal? Acta Univ Cibiniensis 64:40–5
43. O. Toolkit (2011) OECD sustainable manufacturing toolkit, ed: OECD, Publishing, Paris, France, http://www.oecd.org/innovation/green/toolkit

Sustainability Implications of Adopting Industry 4.0 at Different Scales in the Poultry Processing Industry

Derrick Kpakpo Allotey, Ebenezer Miezah Kwofie, and Dongyi Wang

1 Introduction

Industry 4.0 is an advanced manufacturing model that is formed by a set of technologies that makes production systems and industrial activities more integrated, virtual and digital to provide tremendous innovation and competitiveness growth [1, 2]. These technologies include Artificial Intelligence, Robotics, Cyber Physical Systems, Big Data Analytics, Virtual Reality, Augmented Reality, Internet of Things (IoT), Internet of People (IoP) among others [3]. These components are interwoven to produce facets namely Smart Manufacturing, Smart Factory, Smart Supply Chain, Smart Product Development and Smart Life Cycle Analysis [4, 5]. With regards to sustainability, it has been appreciated for its contribution to sustainable manufacturing and environmental management [1, 2, 6]. From the eco-environmental perspective, the use of these technologies helps to optimize the resource usage and reduce the wastage or losses of material flow the hole production and processing streams. [6]. For social sustainability, the smart and autonomous production systems can support employee health and safety, by taking over monotonous and repetitive tasks, which can result in higher employee satisfaction and motivation [1]. However, the technology has not been sufficiently explored from a sustainability perspective due to its novelty and the different degrees of implementation in different production systems to identify possible barriers to sustainability in its adoption [7]. There exist possible negative effects of the Industry 4.0 on sustainability. The metals and other elements used in manufacturing the physical components of the technologies, lead

D. K. Allotey · E. Miezah Kwofie (✉)
Bioresource Engineering, McGill University, 21111 Lakeshore, Sainte Anne-de-Bellevue, QC H9X 3V9, Canada
e-mail: ebenezer.kwofie@mcgill.ca

D. Wang
Department of Biological and Agricultural Engineering, University of Arkansas, AR Fayetteville, USA

H. Gholami et al. (eds.), *Sustainable Manufacturing in Industry 4.0*,
https://doi.org/10.1007/978-981-19-7218-8_8

to land resource usage and environmental pollution, through the mining activities deployed to get them as well as the effect of layoff of labour force with limited technological skills in employing the Industry 4.0 technologies at specific aspects of the production and supply chain [7].

The poultry industry has been one of the thriving sub-sectors of the Food and Agriculture Industry [8]. The poultry industry has a market value of USD 310.7 billion in 2020 and is expected to rise to USD 322.5 billion in 2021 with a compound annual growth rate of 3.8%. The market is expected to reach 422.97 USD billion by 2025 [9]. It is also projected that the global consumption of poultry meat would hit 151.83 kilotonnes by 2030 [10]. This very high consumption rate can be attributed to the highly nutritious and cheap meat and eggs produced from the poultry production and processing industry. Chicken meat (white meat) and eggs do not provide only high-quality protein but also very important and highly required minerals and vitamins. Chicken meat also has less fat (3 g of fat/100 g) as compared to dark (red) meat (5 to 7 g/100 g) [11]. Poultry production is also characterized by the higher conversion rate of feed to meat (2–2.5 kg) in comparison with other livestock like ruminants (red meat) which require about 7 kg of feed to produce 1 kg of meat. Poultry production is also very attractive due to the short production cycle which takes about 7–8 weeks [12].

Traditional poultry processing has been very prevalent in both rural and urban settings. The poultry process line involves the major facet steps: transportation of birds, pre-slaughter, slaughtering, processing (bleeding, scalding, removal of head and feet, evisceration, carcass washing, chilling, deboning, cutting) value addition, utilization of by-products and packaging [13, 14]. Another major challenge faced is inadequate protocols used for poultry meat inspection [15, 16]. There are also problems regarding the processing, storage and cold transportation system for the poultry products. Another major challenge faced is inadequate protocols used for poultry meat inspection [15, 16]. This sometimes causes fluctuations in their supply rates [17]. Also, the repetitive motions, prevailing pay rates and cool temperature of processing plants render these jobs difficult to carry out. There are also high costs associated with training workers to be very precise and fast especially at the cutting section of the processing line [18]. To meet the ever-increasing demand of poultry products as well as addressing the pertinent challenges in the poultry processing industry, companies are adopting new manufacturing technologies. Fortunately, this industry has received enough attention with regard to the application of Industry 4.0 to the different aspects of the processing regarding automated evisceration [19], the use of 3-D imaging for cutting and portioning of the meat [18], smart meat inspection [20], smart meat packaging [21] among others [12, 14, 22–24]. However, there could also be some negative implications of Industry 4.0 to different dimensions of economic, social and environmental sustainability which elicit their evaluation.

This chapter therefore presents an evaluation of the application of Industry 4.0 technologies at different levels of implementation: traditional processing (devoid of Industry 4.0), fully employed Industry 4.0 and a hybrid application of traditional and Industry 4.0 components, under the umbrellas of economic, environmental, and social sustainability. This would highlight the different forms and levels at which the

Industry 4.0 technologies are applied in the poultry processing industry. Moreover, findings from this assessment would inform interested stakeholders in the poultry industry on how to apply these new concepts and the trade-offs associated with adopting the technologies at the different levels of implementation. It would also inform the technology designers of possible sustainability barriers which would be considered in future developments in sustainable manufacturing.

2 Traditional Poultry Processing

Conventional poultry production is an essential part of rural farm household activities; a few birds are reared with little or no feed complement to generate eggs and meat for home consumption and any excess is selling [25].

2.1 Transportation of Bird

Transporting of birds can be very stressful for the birds which usually leads to shrinkage and loss in weight. To avoid this, special modules are used on transport trucks for even air flow and good ventilation [13]. The process flow chart for the poultry processing is summarized in Fig. 1.

2.2 Unloading and Pre-slaughter

Care must be taken when unloading birds to prevent bruises and breaking of bones. At large plants, the broilers are unloaded onto conveyor belts. Feeding is withheld for 8–12 h prior to killing to reduce the amount of feed in the gut and also to prevent tearing during evisceration which can cause faecal contamination to the carcass [13].

2.3 Slaughtering and Processing

2.3.1 Stunning

Usually killing is preceded by stunning. The process of stunning involves dipping the birds in saline water with electric current at relatively low voltages (20 V) to keep them temporarily unconscious [13]. The purpose of this process is to immobilize the birds and make them insensible to the killing process. Stunning also helps to initiate and maximize bleeding [14]. There exist other mechanical and gaseous methods for the stunning process.

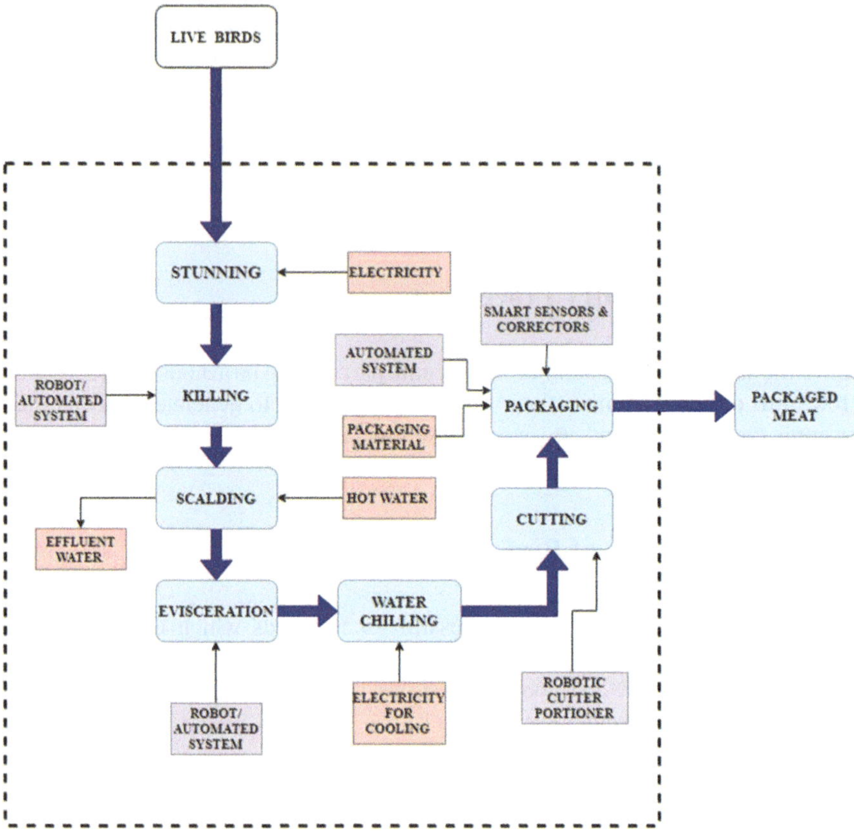

Fig. 1 System boundary for poultry processing (Dusty rose boxes—input/outputs for processes, purple—Industry 4.0 practices)

2.3.2 Killing and Bleeding

In slaughtering the birds, an incision is made to the jugular vein, carotid vein and trachea, or by just above the heart of the birds [23]. This is done by a very sharp knife. In small scale processes, the birds are placed in funnel-shaped kill cones while in largest plants, they are placed on shackles [13]. The efficiency of bleeding has a great effect on the subsequent downstream processes [14]. Maximum bleeding is highly required so there would be no dark spots on the meat [13].

2.3.3 Scalding

The birds are scalded to help loosen the feathers for easy defeathering or picking. Conventionally, hot water baths have been used for scalding but recently steam scalding has been introduced and employed on the large-scale plants [19]. This is done between 50–58 °C [23]. Scalding at high temperatures, to help easy picking, affects the yield in terms of overscalding which 'cooks' the meat and reduces the fat and collagen content. Feathers can be removed by either hand picking or abrasive and rotating actions of rubber or metallic fingers in tub-style pickers [13].

2.3.4 Evisceration

Evisceration is the stage where the internal organs are removed as well as the head and feet [26]. In this process, the dead birds are cut around the vent and the inedible parts such as the small-scale farmers perform evisceration on flat surface while the large plant eviscerate the birds on shackles [13]. The evisceration is performed by supporting the bird with one hand and inserting the fingers of the other hand through the incision in abdomen [19].

2.4 Carcass Processing

2.4.1 Chilling

The carcasses are chilled for a period of time to prevent microbial growth as they await cutting and deboning The most common chilling method are cold water chilling, air chilling and spray chilling [19]. Technically, air chilling is regarded as a cheaper method since there is an absence of water and also a cooling tower, and other ancillary equipment but has higher production cost due to longer times needed for cooling but water chilling method is regarded as more efficient and timesaving. With regard to the meat quality, air chilled meat maintains the natural moisture and colour while absorption of water, as occurs in water chilling, affects the sale price and cooking process due to the added water.

2.4.2 Cutting and Deboning

After chilling, the carcass is typically cut and deboned in different portions to accommodate a variety of different products: drum sticks, thighs, breasts, wings and more [26]. After cutting, the meat can be stored at room temperature for four hours, after which it should be cooled to reduce microbial contamination [12]. Control of the quality of poultry meat includes the following controls: storage space temperature (2–4°C), meat colour, meat elasticity, odour, and meat size or weight according to specifications [12, 19].

2.5 Packaging

Packaging helps to protect the integrity of the processed meat and against microbial contamination. It also helps in easy transportation and retailing of the meat products. There are three types of packaging methods in the poultry processing industry: aerobic, anaerobic and modified atmospheric [14]. Ultrasound applications to meat prior to packaging have also been studied [22]. To maintain customer satisfaction with quality poultry products, each retailer must include an expired information label and treatment when the product is on display [12].

3 Industry 4.0 Advances in Poultry Processing

3.1 Transportation of Birds

Trucks transporting birds from the farms to slaughterhouse are equipped with sensors that monitor environmental properties. They are attached to the cages or modular bins where the birds are loaded. They measure temperature, humidity, ammonia and carbon dioxide levels [27]. Blockchain technologies are also employed in the transportation of the birds to the slaughterhouse For example Transport Genie, a blockchain system, during the transportation of birds from farm to destination, uses blockchain technologies to keep an electronic record of all activities throughout the transportation value chain: loading, transport phase and unloading phase [28]. This track could include every input and output from each link in the supply chain, from breeder, hatchery, producer, producer, slaughterhouse, processor, retailer and consumer [29].

3.2 Automated Killing and Evisceration Lines

High-speed automated bleeding equipment employs a railing system that positions the neck of the suspended birds in such a way that the blood vessels can be opened with precision [19]. Automated systems usually consist of a head puller where an automated guide rail first positions the head into a trough-like structure. While the carcass is moving on the shackle line, the head is pulled back and the neck. The oesophagus and trachea are removed at the same time to save labour [19]. Automated rehanging of carcasses from different shackle lines to the other is labour saving, more hygienic as birds do not touch each other on the sorting table and a more homogenous rigour mortis process [19].

3.3 *Portioning and Cutting*

3.3.1 3-D Imaging for Portioning

The DSI Portioning System has two separate scanning systems. The structural scanning system determines the three-dimensional (3-D) topology of the meat. The cut placement is determined by a 3-D imaging system. When the portioner determines the optimal cuts, an algorithm is executed to cause the cutters to cut and portion the meat at specified loads [18].

3.3.2 Automated Water Jets for Cutting

An automated finely focused water jet is used for cutting (instead of a knife). Speed of cutters (jets) is controlled by the thickness of the meat whereby cutters move relatively slower to ensure a complete cut of very thick meat or parts that contain tendons [18]. There is automated speed control of the cutting system so that the fillets can arrive at the specific desired time for cutting. It can process about 180-200 fillets per min [18].

3.3.3 Fillet-Harvesting Robot

GRIBBOT is a chicken fillet-harvesting robot with 3D vision, a custom-designed gripper, and a transport system to present the breasts to the robotic arm. The GRIBBOT gripper has a curvature and textured surface to better hold the meat while it pulls the fillets from the bone [23]. It mimics the human hand and has a scrap function to reduce meat loss. The robot vision is a visual perception system with a RGB-D camera [24].

3.4 *Automated Hyperspectral-Based Inspection System with Smart Sensors*

Hyperspectral imaging is an emerging smart tool for quality evaluation purposes. It shows a convincing attitude to detect and evaluate chicken meat quality. The method can detect bone fragments as well as faecal contamination [20]. Smart sensors are traditional sensors embedded with intelligence and can perform calculation, conversions and interfaced functions that facilitate self-diagnostics and self-adaptation [4]. After certain calculations and signals, the system can flag any deviation as suspect and these birds are either removed from the line or are more thoroughly inspected. Several systems are already equipped with 'fuzzy-logic' that allows them to 'learn' as new variables are introduced. Robots help in the transfer and rejection of the carcasses [19].

3.5 *Packaging*

3.5.1 Smart Packaging

Freshness indicators and sensor-based devices (biosensors and oxygen sensors) can be integrated into the meat packaging system to monitor the environment of the meat product and provide signals such as temperature, shelf life, spoilage status, etc. [30]. Another dimension is the interactive packaging which is the ability of sensors to detect internal and external changes and to act by changing its own properties. Techniques used to correct change include oxygen scavengers and moisture absorbing systems [21].

3.5.2 Traceability System

Bar code labels and radio frequency identification tags are the most important data carrier devices in the food packaging industry [30]. These are used in conjunction with blockchain technologies where activities at every stage of the value chain are recorded into a large database with encryptions and redistributed among all the links in the chain (slaughterhouse to the plate) has proven to enhance traceability. It helps to build trust among farmers, processing companies and consumers [19].

4 Sustainability Implications for Transition to Industry 4.0

4.1 *Overview and Scenario Selections*

In evaluating the sustainability implications for the transition from traditional poultry systems to more advanced systems where Industry 4.0 is fully implemented, the environmental and economic analyses of three scenarios were considered. The scenarios represent three different levels of implementation of the Industry 4.0 technologies. A first scenario where the product system is based on the traditional process, thus, a normal large plant which operates devoid of Industry 4.0 technologies. In the second scenario, a hybrid system which involves a degree of Industry 4.0 implementation. This hybrid could be considered as semi-Industry 4.0 where some advance manufacturing practices are used. The third scenario where all parts of the process including manufacturing and supply chain fully employ technologies as incorporated in Industry 4.0. A comparative analysis is then carried out to assess these scenarios using the triple bottom line of sustainability-environmental, social and economics as well as product-process safety.

4.2 *Techno-Environmental Impact Assessment*

4.2.1 Method

The techno-environmental assessment of the three scenarios is performed following the life cycle assessment (LCA) methodology. In all the three scenarios, it is assumed that the poultry farm (where the birds are reared), the slaughterhouse and the carcass processing plant are at the same location hence the environmental impact due to transportation activities are excluded. The functional unit for LCA is 1 kg of packaged meat. This contains life cycle inventory where quantitative data on the different inputs and outputs defining the individual process are presented. This data was mainly sourced from available literature and Ecoinvent Database. The environmental impacts resulting from the inputs are evaluated through a life cycle analysis conducted in the OpenLCA software. Impact categories studied were Global Warming Potential (GWP), Abiotic Depletion Potential (ADP) and Cumulative Energy Demand (CED). The life cycle inventory included the environmental impact associated with the robot production of the robot.

4.2.2 Results

LCI from the different processes as by the scenarios are presented in Table 1. Environmental impacts resulting from the LCA are also presented in Fig. 2. Some processes were excluded from the LCI since there was no quantitative flow of input and output. The LCI does not incorporate data on how robots and automated systems offset or reduce the resource and emissions on the processing plant since there is limited quantitative data describing these offsets. But it contains the LCI on the production of the cyber physical system.

Table 1 LCI on the Different Scenarios per functional unit

	S1	S2	S3
Inputs			
Water, kg	8.4	10	1
Fuel (Natural gas), kg	0.072	1.02	2.3
Electricity, kWh	0.13	2.6	3.8
Outputs			
Methane, mg	7.51	17.6	20.6
Carbon dioxide (fossil), g	55.1	75	85
Carbon mono-oxide, mg	10.1	32	38
Nitrogen dioxide, mg	75.01	88.2	100.1
Sulphur dioxide, mg	105	127	142

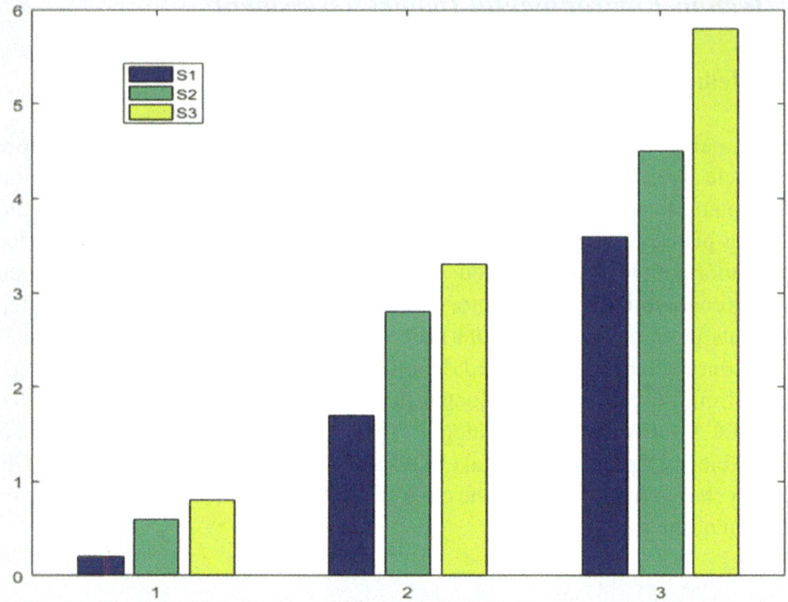

Fig. 2 Environmental impact results from the three scenarios. 1(GWP-kg CO2), 2 (g Sb eq ADP), 3(CED-MJ eq)

4.3 Techno-Economic Sustainability Analysis

4.3.1 Method

Techno-economic indicators employed for the evaluation were Total Capital Cost, Production Cost, Process Capacity and Process Loss (Meat loss). The Total Capital Cost includes the total cost of the equipment (purchasing and installation. The Production cost entails cost of birds (raw material) and utilities (electricity, water, fuel, packaging) as well as labour costs. Production Capacity is defined as the total sales (or revenue) which is determined as the product of the production rate (how much meat produced per day) and the selling price of 1 kg of chicken meat (priced at $5per kg). Meat loss is defined as the cost of chicken meat (or pieces) that are lost along the various parts of the processing line.

4.3.2 Results

Results from the economics are summarized in Table 2.

Table 2 Economic performance of the three implementation scenarios

Indicators	S1	S2	S3
Total capital cost, $	80,000	280,000	500,000
Production cost, $/kg of meat	2.5	11	17
Production capacity, $	50,000	90,000	150,000
Meat loss, $	300	150	50

4.4 Discussion

Sustainable Manufacturing 4.0 does not only involve how Industry 4.0 tools help to reduce environmental, economic and social burden associated with processing systems but also should involve the sustainable manufacturing of the technologies tools used to achieve this. From the environmental impact results, the fully automated system posed the greatest danger to the environment. This is not expected since Industry 4.0 practices are highly acclaimed for its contribution to a sustainable environment through advanced technologies to minimize emissions and resource usage.

However, the inclusion of environmental impact associated with the manufacturing of the robots and other cyber physical systems, in the sustainability evaluations, has received little attention. The high energy demand required to smelt and mould metals, steel, aluminium, etc. relays into a proportionally high cumulative energy demand (CED). The burning of fuel needed to produce the high heat energy also results in high emission of gases that contributes to Global Warming Potential (GWP). The mining activities carried out to obtain the raw metals and ores for the manufacturing of the robots also pose a heavy burden on the land (resource depletion) on par with the air pollution associated with these activities.

Sustainability evaluation studies can be channelled into the cyber physical systems production to identify hotspots needed to reduce the environmental burden associated with the processes. On the other hand, more surveys and research studies can be conducted to quantify the extent at which the Industry 4.0. practices reduce environmental impact when incorporated into the poultry processing lines as well other production lines in the food and agriculture industry. On the pathway to achieving Sustainable Manufacturing 4.0, this could then be compared to environmental impact of the technology production system to determine the balance and the facet which needs to be improved.

From the economic analysis (Table 2), capital cost and production costs are highest in S2 and S3. This is mainly because of the very high cost of the industrial robots, averagely USD 50,000 which is also due to both the expensive materials used for their construction, (the metals and electronic elements) and the highly powerful software embedded in them. This is the same as the automation systems. The high electric power demand, averagely 8 kW, of these robots and automated systems contribute significantly to the total production costs. Studies can be conducted to specifically determine the offsets in labour cost against the operating costs in semi-automated and fully automated poultry processing plants.

The loss in the poultry process factory diminishes sharply from S1 to S3. The better optimization by incorporating blockchain technologies and the better precision of the cyber physical systems as the full Industry 4.0 practices helps to minimize resource usage and loss at the slaughterhouse and processing plants. The results also show that there are always trade-offs regarding the implementation of the Industry 4.0 practices in the capital cost, production cost and production capacity. Since there are differences in trends between the environmental and economic impacts some decisions of implementation level affect the economic performance (negatively or positively), the converse can occur in the environmental impact.

5 Conclusion

The available technologies such as blockchain, big data analytics, cyber physical systems, IoT, etc. in Industry 4.0, which produce a tapestry of interconnections between humans (labour) and technologies, have been applied in diverse forms of manufacturing and production systems, with the poultry processing industry not left out. Sustainability has also been a great concern as industries strive to meet its demanding goals while still maximizing their profits. On the pathway to achieve Sustainable Manufacturing with adoption of Industry 4.0 technologies, three scenarios were created with different implementation levels, a traditional process, semi-automated and fully automated systems. Environmental and economic impact assessments were conducted for the three scenarios Life Cycle Analysis (LCA) was employed for the environmental impact assessment. Impact categories studied were Global Warming Potential, Results from the LCA showed an unexpected trend in the impact results. Thus, the semi and fully automated systems posed higher env. Economic performance of the system. Industry 4.0 is very promising in contributing to sustainable manufacturing in terms of increasing production capacity, minimizing loss and optimizing processes. However, a careful study on how to make the production of the cyber physical systems (robots) more sustainable and quantitively defining the offsets in resource usage and emissions associated with it application would go along to enhance sustainable poultry processing as well as the manufacturing industry as a whole. Tradeoffs among individual economic and environmental impacts with different Industry 4.0 adoption levels would also contribute massively to this new revolution.

References

1. Bai C, Dallasega P, Orzes G, Sarkis J (2020) Industry 4.0 technologies assessment: a sustainability perspective. Int J Prod Econ 229:1–15
2. Carvalho ACP, Carvalho APP, Carvalho NGP (2020) Industry 4.0 technologies: what is your potential for environmental management? In: Industry 4.0—current status and future trends
3. Massaro M, Secinaro S, Mas FD, Brescia V, Calandra D (2021) Industry 4.0 and circular economy: an exploratory analysis of academic and practitioners' perspectives. Bus Strat Env. 2021:1213–1231
4. Cimini C, Pezzotta G, Pinto R, Cavalieri S (2019) Industry 4.0 technologies impacts in the manufacturing and supply chain landscape: an overview In: Proceedings of SOHOMA 2018 Industry 4.0 technologies impacts in the manufacturing and supply chain landscape: an overview, pp 1–12
5. Corallo A, Latino ME, Menegoli M (2018) From Industry 4.0 to Agriculture 4.0: a framework to manage product data in agri-food supply chain for voluntary traceability. Int J Nutr Food Eng 12
6. Ejsmont K, Gladysz B, Kluczek A (2020) Impact of Industry 4.0 on sustainability—bibliometric literature review. Sustainability 12:1–29
7. Bonilla SH, Silva HRO, Silva MTD, Gonçalves RF, Sacomano JB (2018) Industry 4.0 and sustainability implications: a scenario-based analysis of the impacts and challenges. Sustainability 10:3740
8. Research & Markets (2021) Global organic poultry market report 2021
9. Yıldız D (2021) Global poultry industry and trends. Int Mag Animal Feed Additives Industry
10. Shabendeh M (2021) Global poultry meat consumption 2021–2030.
11. Farrell D (2013) The role of poultry in human nutrition. In: Poultry development review. FAO, pp 1–8
12. Wahyono ND, Utami MMD (2018) A review of the poultry meat production industry for food safety in Indonesia. In: The 2nd international joint conference on science and technology (IJCST) 2017 (vol 953). Indonesia
13. Fanatico A (2003) Small-scale poultry processing. In: National center for appropriate technology (NCAT)
14. ICAR (2021) Handbook of meat and poultry processing. ICAR National Research Centre on Meat
15. Haligah S (2017) Analyzing production challenges plaguing Ghana's agricultural sector: a focus on poultry production in Greater Accra. Ashesi University College
16. Sonaiya EB, Swan SEJ (2004) Small-scale poultry production. In: Animal production and health, technical guide. FAO, p 120
17. Mitra A, Majumder D, Mishra M, Sarkar S (2021) Poultry farming: prospects and impediments in India. Saudi J Humanit Soc Sci 6:193–198
18. Heck B (2006) Automated chicken processing. IEEE Control Syst Mag:17–19
19. Barbut S (2015) Primary processing of poultry. In: The science of poultry and meat processing
20. Xiong Z, Xie A, Sun DW, Zeng XA, Liu D (2014) Application of hyperspectral imaging in chicken meat safety and quality detection and evaluation: a review. Crit Rev Food Sci Nutr:55
21. SGCS (2019) Packaging methods used in poultry Industry. In: Poultry reporter
22. Al-Hilphy AR, Al-Temimi AB, Al Rubaiy HHM, Anand U, Delgado-Pando G, Lakhssassi N (2020) Ultrasound applications in poultry meat processing: a systematic review. J Food Sci 85:1386–1396
23. Esper IDM, From PJ, Mason A (2021) Robotisation-and-intelligent-systems-in-a_2021_Trends-in-Food-Science-Tech.pdf. Trends Food Sci Technol:214–222
24. Misimi E, Øye ER, Eilertsen A, Mathiassen JR, Asebø OB, Gjer-stad T, Buljo J, Skotheim Ø (2016) Gribbot-Robotic 3D vision-guided harvesting of chicken fillets. Comput Electron Agric 121:84–100
25. Ali MM, Hossain MM (2012) Problems and prospects of poultry industry in Bangladesh: an analysis. In: AIUB Bus Econ Working Paper Series, vol 01

26. NCC (2020) Processing: how are chickens slaughtered and processed for meat. In: The poultry processing line
27. Esnaola-Gonzalez I, Gomez-Omella M, Ferreiro S, Fernandez I, Lazaro I, Garcia E (2020) An IoT platform towards the enhancement of poultry production chains. Sensors (Basel):20
28. Thornton G (2019) 3 Poultry technology innovations to watch. In: Wattpoultry
29. Baerg M (2019) Blockchain for poultry has arrived. In: Canadian poultry
30. Chowdhurry EU, Morey A (2019) Intelligent packaging for poultry industry. J Appl Poultry Res 28:791–800

Horizontal Collaboration Business Model Towards a Sustainable I4.0 Value Creation

Georgina Elizabeth Riosvelasco-Monroy, Iván Juan Carlos Pérez-Olguín, Julieta Flores-Amador, Luis Asunción Pérez-Domínguez, and Jesús Andrés Hernández-Gómez

1 Introduction

Sustainable Manufacturing 4.0 establishes disruptive events toward business model creation; through Industry 4.0 elements. Horizontal collaboration seen as one of I4.0 elements, integrates within a value creation network; seen as an alliance where a common goal is set, addressing opportunities for supply chain integration. Being laid out as the new alternative from vertical collaboration, horizontal collaboration is a strategic move in which enterprises can gain competitive advantage, knowledge transfer, information sharing, information technology, economies of scale, among other benefits. In order for horizontal collaboration can become a long-term value creation network, enterprises must effectively carry out certain factors defined by the alliance, to stimulate common goals such as a sustainable manufacturing 4.0.

In this chapter, we present a new understanding of horizontal collaboration as a value creation network through a business model integration to influence collaborative projects seeking for a position within a supply chain and enhancing sustainable manufacturing 4.0 development. This chapter is organized into five sections: Sect. 2 an introduction to supply chain horizontal collaboration; Sect. 3 a review of business model components as a framework of an enterprise's structure as an integration of

G. E. Riosvelasco-Monroy (✉) · I. J. C. Pérez-Olguín (✉) · L. A. Pérez-Domínguez · J. A. Hernández-Gómez
Department of Industrial and Manufacturing Engineering, Institute of Engineering and Technology, Universidad Autonoma de Ciudad Juarez, Av. Del Charro, 32584 Cd. Juarez, Chihuahua, Mexico
e-mail: georgina.riosvelasco@uacj.mx

I. J. C. Pérez-Olguín
e-mail: ivan.perez@uacj.mx

J. Flores-Amador
Institute of Social Science and Administration, Universidad Autonoma de Ciudad Juarez, Av. Heroico Colegio Militar, 32000, Cd. Juarez, Chihuahua, Mexico

H. Gholami et al. (eds.), *Sustainable Manufacturing in Industry 4.0*,
https://doi.org/10.1007/978-981-19-7218-8_9

horizontal collaboration factors; Sects. 4 and 5 a research framework of multi-criteria decision-making techniques presenting a CODAS-HTFLS-Mahalanobis approach to identify horizontal collaboration top factors grouped within the business model components. Section 6 displays COHRV Model for sustainable manufacturing 4.0 environment.

2 Supply Chain Horizontal Collaboration

Collaboration is defined as a status when two or more enterprises work on a joint venture or alliance where a common goal is set, addressing opportunities for supply chain integration; specifically, a growing involvement for enterprises [1, 2]. A collaborative supply chain is a cooperation and business philosophy between enterprises that share tangible and intangible assets, and joint tasks [3, 4, 5]. Introducing, from literature, three types of collaboration: (1) vertical, joint ventures between customers, suppliers, and the enterprise; (2) horizontal, enterprises from the same echelon working together sharing information and resources; and (3) lateral, benefiting from both alliances, vertical and horizontal [2, 6, 7].

Horizontal collaboration exposes an increment in academic interest, being laid out as the new alternative from the vertical collaboration [8]. Defined as a (1) strategy that eases the joint venture of techniques and processes, allowing enterprises to add value within their supply chain for creating and developing new technology [9, 10]. Presenting characteristics such as creating a new organizational structure by making available intellectual and machinery assets (Trejo et al. 2010). A strategic move in which enterprises can gain competitive advantage, knowledge transfer, information sharing, information technology, and economies of scale, among other benefits.

Even though it lays down its beneficial side, horizontal collaboration lacks a model or methodology for an effective implementation thus reducing an already higher percentage at initiative failure [11]. In order for horizontal collaboration to become a long-term partnership and a value creation network, enterprises must effectively carry out strategies developed through a horizontal collaboration business model.

3 Business Model Research Framework

A business model is well known as a visual framework of each enterprise's internal and external structure, which enhances the essential elements that consider each activity such as value proposition, available resources, supply chain structure, and market necessities [12, 13, 14, 15]. From this perspective, a horizontal collaboration framework is designed and proposed through a business model. To reach this objective, horizontal collaboration factors are grouped within three business model components (1) Content, what activities are selected from the alliance, (2) Structure, how does these activities are linked; and (3) Governance, who will be responsible for the compliance of the activities and strategies developed through this business model.

3.1 Content Component

The enterprise content involves those activities that satisfy customers' necessities, as well as its service and profitability [16], aligning each enterprises' strategy [17]. Table 1 lists the horizontal collaboration factors regarding the *what* content of the collaboration.

3.2 Structure Component

Considering a series of points for structuring each activity, such as prioritizing and handling with neutrality the structure, implementing the right IT interfaces [20]. Table 2 shows the horizontal collaboration factors that take part in the *how* to structure and link the collaboration.

3.3 Governance Component

Each enterprise activity must have a design network of roles and responsibilities for the newly defined activities, ensuring transparency [16, 17]. Table 3 presents the list of horizontal collaboration factors related to the *who* in the collaboration.

4 Multi-criteria Decision-Making Techniques (MCDM)

MCDM techniques have been used by researches in the academic field to evaluate a set of alternatives from a variety of criteria [40, 41]. Diverse MCDM techniques have been suggested for satisfying different conditions, such as supplier selection [42], identifying critical factors for organizational culture in innovation [43], and site selection for a desalination plant [44].

A decision matrix is presented as a method for an MCDM problem which shows the evaluation values of each variable with respect to each criterion. Table 4 shows the MCDM decision matrix for m HC factors $X = \{x_1, x_2, \ldots, x_m\}$ evaluated on a finite set of n criterion $C = \{C_1, C_2, \ldots, C_n\}$.

Suggesting a series of MCDM techniques for determining horizontal collaboration factors hierarchical analysis. The specific techniques are as follows (1) HFLTS, which delimits the values that an expert panel will use for each evaluation; (2) AR,

Table 1 Content component items from horizontal collaboration factors

Code	Content component items	References
C01	Logistics	[18, 19, 20, 21, 22]
C02	Delivery performance	[18, 21, 22, 23]
C03	Marketing and sales events	[18]
C04	Complementary assets	[18, 24]
C05	Process performance	[18, 25, 26]
C06	Problem solving and support	[18]
C07	Paperwork and administrative process support	[18]
C08	Production flexibility	[21, 24, 26, 27, 28, 29, 30, 31]
C09	Focus strategy in limited resources	[32]
C10	Information sharing	[9, 11, 19, 23, 25, 28, 29, 31, 33, 34, 35, 36]
C11	Trust	[11, 19, 21, 22, 28, 29, 30, 31, 34, 35, 36]
C12	Performance metrics	[2, 37]
C13	Incentive alignment	[23, 31, 33]
C14	Channel alignment	[25]
C15	Operational performance	[25, 28, 29, 30, 38]
C16	Negotiation performance	[11, 20, 22, 23, 28, 31, 33, 35, 38]
C17	IT implementation	[20, 22, 23, 38]
C18	Technology implementation	[9, 24, 30, 39]
C19	Forecasting	[25, 27]
C20	Waste reduction	[25]
C21	Resources sharing	[11, 21, 29, 34]
C22	Knowledge transfer	[9, 28, 29, 34, 39]
C23	Goal alignment	[9, 28, 29, 35]
C24	Research & Development	[9, 35]
C25	Enterprise performance	[21, 28, 35]
C26	Initiation phase	[39]
C27	Development phase	[39]
C28	Commercialization phase	[39]
C29	Value creation	[21]
C30	Joint purchases	[38]

a technique that incorporates the acquired knowledge from the literature review and the AHP decision; and (3) CODAS, a combinative technique that uses two sets of distance measure to determine the hierarchy from the factors evaluated, where the authors propose, the use of the Mahalanobis distance measure instead of the Taxicab distance for the secondary measure set.

Table 2 Structure component items from horizontal collaboration factors

Code	Structure component items	References
E01	Marketing and sales events	[18]
E02	Word of mouth publicity	[18]
E03	Economies of scale	[18, 31, 32]
E04	Process performance	[18, 25, 26]
E05	Problem solving and support	[18]
E06	Paperwork and administrative process support	[18]
E07	Costs	[26]
E08	Innovation	[9, 21, 26, 28, 34, 35]
E09	Quality	[26, 34]
E10	Sustainability	[26]
E11	Cluster strategy	[32]
E12	Focus strategy in limited resources	[32]
E13	Supply chain integration	[23, 23]
E14	Information sharing	[9, 11, 19, 23, 25, 28, 29, 31, 33, 34, 35, 36]
E15	Customer performance	[21, 23, 28]
E16	Common interest	[11, 36, 38]
E17	Openness	[9, 11, 29, 36]
E18	Trust	[11, 19, 21, 22, 28, 29, 30, 31, 34, 35, 36]
E19	Cooperation	[36]
E20	Leadership	[36]
E21	Decision synchronization	[29, 33, 34]
E22	Operational performance	[25, 28, 29, 30, 38]
E23	Organizational structure compatibility	[9, 19, 28, 34, 38]
E24	IT implementation	[20, 22, 23, 38]
E25	Distribution centres and warehouses compatibility	[19, 24, 31]
E26	Technology implementation	[9, 24, 30, 39]
E27	Geographic area compatibility	[20, 22, 27, 33]
E28	Business strategy	[34]
E29	Communication	[11, 30, 31, 34]
E30	Research & Development	[9, 35]
E31	Enterprise performance	[21, 28, 35]
E32	New Product Development phases	[39]
E33	Transaction cost analysis	[39]

(continued)

Table 2 (continued)

Code	Structure component items	References
E34	Market strategy	[21, 27, 28, 39]
E35	Networking	[20, 21, 27, 28]
E36	Joint purchases	[38]

4.1 Criterion Weight

Introduced by [43] AR is used as a technique for reducing the ambiguity of the values obtained from an expert panel. This, by incorporating the weighted values determined by the average of the following analysis (1) AK, the weighted value of the frequency of variables identified through a literature review; and, (2) AHP, the weighted value as a result of a complex hierarchy decision. Using $w_j^{AR} = \gamma w_j^{AK} + (1 - \gamma) w_j^{AHP}$; where γ represents the impact from the dimensional criterion weighting that will have of respect to the decision makers; w_j^{AK} is the obtained weighting from the literature review for the critical dimension j; w_j^{AHP} is the AHP weighting for the critical criterion j; and finally, w_j^{AR} is the ambiguity reduction weighting for the critical criterion j.

The above criterion weight proposal integrates n documented research provided by experts, additional to the expert panel that participated in the weighting values in this paper.

4.2 Analytic Hierarchical Process (AHP)

Created by Thomas Saaty in the 1970s, AHP is recognized as an MCDM methodology that deals with multi-criteria ranking and a set of tangible and intangible alternatives [45]. Its importance lies in the capability to structure hierarchically each complex, multi-person, multi-criteria problem throughout a construction of a pairwise comparison matrix [58]. It considers three main components, such as (1) decomposing the elements of the complex problem into a hierarchy, facilitating the decision makers to identify the major components in an efficient way, (2) providing a measurement methodology called pairwise comparison matrix, an off-diagonal relationship of one side and placing the reciprocals in the transposed positions, for establishing priorities among the elements; and, (3) calculating the priorities and consistency of the data provided by the expert panel.

This matrix uses a 9-point scale, also known as Saaty judgement scale [58] and the main axioms considered are [43]:

1. Reciprocal judgments, where A is a matrix of paired comparisons $a_{ij} = 1/a_{ij}$.
2. Condition of homogeneity of the elements, the elements are compared in the same order of magnitude.

Table 3 Governance component items from horizontal collaboration factors

Code	Governance component item	References
G01	Delivery performance	[18, 21, 22, 23]
G02	Marketing and sales events	[18]
G03	Word of mouth publicity	[18]
G04	Economies of scale	[18, 31, 32]
G05	Process performance	[18, 25, 26]
G06	Problem solving and support	[18]
G07	Paperwork and administrative process support	[18]
G08	Production flexibility	[21, 24, 26, 27, 28, 29, 30, 31]
G09	Innovation	[9, 21, 26, 28, 34, 35]
G10	Quality	[26, 34]
G11	Sustainability	[26]
G12	Cluster strategy	[32]
G13	Focus strategy in limited resources	[32]
G14	Multidisciplinary team	[11, 34]
G15	Information sharing	[9, 11, 19, 23, 25, 28, 29, 31, 33, 34, 35, 36]
G16	Customer performance	[21, 23, 28]
G17	Common interest	[11, 36, 38]
G18	Openness	[9, 11, 29, 36]
G19	Trust	[11, 19, 21, 22, 28, 29, 30, 31, 34, 35, 36]
G20	Cooperation	[36]
G21	Activities prioritization	[36]
G22	Leadership	[36]
G23	Performance metrics	[37]
G24	Decision synchronization	[29, 33, 34]
G25	Incentive alignment	[23, 31, 33]
G26	Partner selection	[11, 19, 25, 27, 38]
G27	Confidence	[38]
G28	Commitment	[31, 34, 38]
G29	Organizational structure compatibility	[9, 19, 28, 34, 38]
G30	Negotiation performance	[11, 20, 22, 23, 28, 31, 33, 35, 38]
G31	Outsourcing	[24]
G32	Geographic area compatibility	[20, 22, 27, 33]
G33	Opportunity	[28, 33]
G34	Confidentiality	[20, 33]
G35	Business strategy	[34]

(continued)

Table 3 (continued)

Code	Governance component item	References
G36	Communication	[11, 30, 31, 34]
G37	Resources sharing	[11, 21, 29, 34]
G38	Knowledge transfer	[9, 28, 29, 34, 39]
G39	Absorption capacity	[9]
G40	Goal alignment	[9, 20, 29, 35]
G41	Business model identification	[35]
G42	Enterprise performance	[21, 28, 35]
G43	Market strategy	[21, 27, 28, 39]
G44	Stock market reaction	[39]
G45	Networking	[20, 21, 27, 28]
G46	Stakeholder identification	[11]

Table 4 Multi-criteria decision matrix

Criterion						
Factors	X_i / C_j	C_1	C_2	C_3	…	C_n
	X_1	X_{11}	X_{12}	X_{13}	…	X_{1n}
	X_2	X_{21}	X_{22}	X_{23}	…	X_{2n}
	X_3	X_{31}	X_{32}	X_{33}	…	X_{3n}
	⋮	⋮	⋮	⋮		⋮
	X_m	X_{m1}	X_{m2}	X_{m3}	…	X_{mn}

3. Condition of hierarchical structure or reuse dependent.
4. Condition of rank order expectations, which are structured in alternatives and criteria.

4.3 Hesitant Fuzzy Linguistic Sets (HFTLS)

HFTLS is presented as a tool for experts to deliver their assessments using linguistic expressions [46]. Suggesting the use of HFTLS to determine the hierarchy of horizontal collaboration factors within three business model concepts. The advantage that this tool presents as a MCDM technique, is that it gives the expert panel the requirements when there are doubts present in a qualitative context. The basic terms and operations for HFTLS application are as follows:

Definition 1 Let T be a linguistic term set, $T = \{T_0, \ldots, T_i\}$; an HFLTS, HT, is an ordered finite subset of the consecutive linguistic term of T. When $H_T(\tau) = \{\}$, the

HFLTS is called an empty set; in the case of $H_T(\tau) = T$ the set is denominated a full HFLTS, and when $H_T(\tau) = \{\mu : \mu \subseteq \tau\}$, μ is a subset of T.

Definition 2 Let T be a linguistic term set, $T = \{T_0, \ldots, T_i\}$, and HT, H_T^1 and H_T^2 be the three HFTLS. The HT + (upper bound) and HT- (lower bound) are defined as $H_{T+} = max(t_i) = t_j, t_i \epsilon H_T \, and \, t_i \leq t_j \forall i$ and $H_{T-} = min(t_i) = t_j, t_i \epsilon H_T \, and \, t_i \geq t_j \forall i$.

Definition 3 The complement of an HFTLS, H_T, is defined as $H_T^c = T - H_T = \{t_i : t_i \in T \, and \, t_i \notin H_T\}$. In addition, the evolutive complement of H_T is $\left(H_T^c\right)^c = H_T$, due $H_T^c = T - H_T$ then $\left(H_T^c\right)^c = T - H_T^c = T - (T - H_T) = H_T$.

Definition 4 The union between $H_T^1 \, and \, H_T^2$ is defined as $H_T^1 \cup H_T^2 = \left\{t_i / t_i \in H_T^1 or t_i \in H_T^2\right\}$. In other words, the union of two HFTLS is the set of elements included in both $H_T^1 \, and \, H_T^2$.

Definition 5 The intersection between $H_T^1 \, and \, H_T^2$ is defined as $H_T^1 \cap H_T^2 = \left\{t_i / t_i \in H_T^1 \, and \, t_i \in H_T^2\right\}$. This means that the intersection of two HFTLS is the set that contains the elements included in H_T^1 and also included in H_T^2.

Definition 6 The linguistic interval with upper bound and lower bound limits obtained from maximum and minimum linguistic terms are called envelope of HFTLS, $Env(H_T)$, and is defined as $Env(\mathrm{H_T}) = \left[\mathrm{H_{T+}}, \mathrm{H_{T-}}\right]$.

Definition 7 The comparison between $H_T^1 and H_T^2$ is defined as $H_T^1(\tau) > H_T^2(\tau) if \, Env(H_T^1(\tau)) > Env\left(H_T^2(\tau)\right)$ and $H_T^1(\tau) = H_T^2(\tau) if \, Env(H_T^1(\tau)) = Env\left(H_T^2(\tau)\right)$.

4.4 Combinative Distance-Based Assessment (CODAS)

CODAS is presented as a decision-making methodology for MCDM problems, which enhances features not performed by other MCDM techniques [40]. The main objective of this method is to determine the hierarchical alternatives by using two distance measures. Used in different situations, CODAS is proposed as an alternative to select the desire industrial robot from a five criteria and m alternatives matrix [40], or, developed as a CODAS analysis to enhanced organizational culture factors through an innovation in Industry 4.0 [43],and, a selection of the best location for a desalination plant in Libya [44].

CODAS uses two distance methods, being the Euclidean distance as the main distance measure and, the Taxicab distance as the secondary distance measure, calculated by the negative ideal distance [40]. Resulting the desirable alternative, the one that presents a greater distance from the negative ideal solution [40, 43, 44]. From this point forward, the authors address a CODAS methodology by evaluating different distances, particularly Mahalanobis distance.

Presenting the steps for developing the CODAS method, modifying the original terms and the way to rank the preference score [40]:

Step 1. Construction of the decision matrix.

$$T = [T]_{n\times m} = \begin{bmatrix} t_{11} & t_{12} & \cdots & t_{1m} \\ t_{21} & t_{22} & \cdots & t_{2m} \\ \vdots & \vdots & \vdots & \vdots \\ t_{n1} & t_{n2} & \cdots & t_{nm} \end{bmatrix} \quad (1)$$

where T_{ij}, shows the value of the i alternative in the criterion j, $i \in \{1, 2, \ldots, n\}$ and $i \in \{1, 2, \ldots, m\}$.

Step 2. Calculating the normalized decision matrix.

$$n_{ij} = \begin{cases} \frac{t_{ij}}{\max\limits_i t_{ij}} if\ j \epsilon N_b \\ \frac{\min\limits_i t_{ij}}{t_{ij}} if\ j \epsilon N_c \end{cases} \quad (2)$$

where N_b y N_c are a set of significant dimensional criteria.

Step 3. Calculating the normalized weight in the decision matrix by $r_{ij} = w_j t_{ij}$. Where w_j is the criterion's weight value j, with $0 < w_j < 1$ and $\sum_{j=1}^{m} w_j = 1$.

Step 4. Defining the ideal negative solution with $ns = ns_{j1xm}$ and $ns_j = \min\limits_i r_{ij}$.

Step 5. Calculate the main distance and secondary distance of the negative idea solution alternatives.

Step 6. Grounding the relative evaluation matrix as $R_a = h_{iknxn}$ and $h_{ik} = (E_i - E_k) + (\varphi(E_i - E_k) \times (T_i - T_k))$. Where $i \in \{1, 2, \ldots, n\}$ and φ shows a threshold function to recognize the equality of the distances of the two alternatives defined by:

$$\varphi(x) = \begin{cases} 1 & \text{if } |x| \geq r \\ 0 & \text{if } |x| < r \end{cases} \quad (3)$$

r value is set by the decision makers, within a range parameter of 0.01 and 0.05. For the purposes of the present study $r = 0.03$.

Step 7. Defining the score from each evaluated alternative, through $T_i = \sum_{k=1}^{n} t_{ik}$.

Step 8. Rank the alternatives by T_i score value.

For this proposal, the T_i score value is being selected with the lowest value as the best option among the alternatives. Due to the fact, that by observing the research results, this criterion presents a ranking aligned to the expert panel's decision-making matrix evaluation aggregated to the preference values (T_i) obtained for each alternative. If used the inverse criterion, the first chosen alternatives within the ranking are the worst alternatives evaluated by the expert panel.

4.5 Distance Measures Between Two Sets

Continuing with the above definitions, this section keeps going on to Definition 8. Introducing a new combined distance measure set using Mahalanobis distance as a secondary measure. The following definitions demonstrate the distance measures from two collections of HFTLS [47].

Definition 8 Let T be a linguistic term set, $T = \{T_\alpha : \alpha = -\tau, \ldots, \tau\}$, H_T^1 and H_T^2 be two HFLTS. The similarity measure is defined as $\rho(H_T^1, H_T^2)$ were $0 \le \rho(H_T^1, H_T^2) \le 1$; $\rho(H_T^1, H_T^2) = 1 \Leftrightarrow (H_T^1 = H_T^2)$; $\rho(H_T^1, H_T^2) = \rho(H_T^2, H_T^1)$.

Definition 9 Let T be a linguistic term set, $T = \{T_\alpha : \alpha = -\tau, \ldots, \tau\}$, H_T^1 and H_T^2 be two HFLTS. The relationship between distance and similarity measure is defined as $\rho(H_T^1, H_T^2) = 1 - d(H_T^1, H_T^2)$.

Definition 10 Let T be a linguistic term set, $T = \{T_\alpha : \alpha = -\tau, \ldots, \tau\}$, of a HFLTS. The linguistic terms, $T_\alpha, T_\beta \in T$, the distance measure between T_α and T_β is defined as $d(T_\alpha, T_\beta) = |\alpha - \beta|/(2\tau + 1)$. Where $2\tau + 1$ is the number of linguistic terms in the set T.

Definition 11 Let T be a linguistic term set, $T = \{T_\alpha : \alpha = -\tau, \ldots, \tau\}$, H_T^1 and H_T^2 be two HFLTS. The Hamming distance of $H_T^1(x_i)$ and $H_T^2(x_i)$ can be defined as:

$$D_1\left(H_T^1(x_i), H_T^2(x_i)\right) = \frac{1}{L}\sum_{l=1}^{L}\frac{\left|\delta_l^1 - \delta_l^2\right|}{2\tau + 1} \tag{4}$$

Definition 12 Let T be a linguistic term set, $T = \{T_\alpha : \alpha = -\tau, \ldots, \tau\}$, H_T^1 and H_T^2 be two HFLTS. The Hamming-Euclidean distance of $H_T^1(x_i)$ and $H_T^2(x_i)$ can be defined as:

$$D_2\left(H_T^1(x_i), H_T^2(x_i)\right) = \left(\frac{1}{T}\sum_{l=1}^{T}\left(\frac{\left|\delta_l^1 - \delta_l^2\right|}{2\tau + 1}\right)^2\right)^{1/2} \tag{5}$$

Definition 13 Let T be a linguistic term set, $T = \{T_\alpha : \alpha = -\tau, \ldots, \tau\}$, H_T^1 and H_T^2 be two HFLTS. The generalized Hausdorff distance of $H_T^1(x_i)$ and $H_T^2(x_i)$ can be defined as:

$$D_3\left(H_T^1(x_i), H_T^2(x_i)\right) = \left(\max_{l=1,2,\ldots,T}\left(\frac{\left|\delta_l^1 - \delta_l^2\right|}{2\tau + 1}\right)^\lambda\right)^{1/\lambda} \tag{6}$$

where $\lambda > 0$, for $\lambda = 1$ the generalized Hausdorff distance becomes the Hamming-Hausdorff distance, for $\lambda = 2$ the generalized Hausdorff distance becomes the Euclidean-Hausdorff distance.

Definition 14 Let T be a linguistic term set, $T = \{T_\alpha : \alpha = -\tau, \ldots, \tau\}$, H_T^1 and H_T^2 be two HFLTS. The hybrid Hamming-Hausdorff distance of $H_T^1(x_i)$ and $H_T^2(x_i)$ can be defined as:

$$\boldsymbol{D}_4\left(\boldsymbol{H}_T^1(\boldsymbol{x}_i), \boldsymbol{H}_T^2(\boldsymbol{x}_i)\right) = \frac{1}{2}\left(\frac{1}{\boldsymbol{T}}\sum_{l=1}^{T}\frac{\left|\delta_l^1 - \delta_l^2\right|}{2\tau + 1} + \max_{l=1,2,\ldots,T}\left(\frac{\left|\delta_l^1 - \delta_l^2\right|}{2\tau + 1}\right)\right) \quad (7)$$

Definition 15 Let T be a linguistic term set, $T = \{T_\alpha : \alpha = -\tau, \ldots, \tau\}$, H_T^1 and H_T^2 be two HFLTS. The hybrid Euclidean Hamming-Hausdorff distance of $H_T^1(x_i)$ and $H_T^2(x_i)$ can be defined as:

$$\boldsymbol{D}_5\left(\boldsymbol{H}_T^1(\boldsymbol{x}_i), \boldsymbol{H}_T^2(\boldsymbol{x}_i)\right) = \frac{1}{2}\left(\frac{1}{\boldsymbol{T}}\sum_{l=1}^{T}\left(\frac{\left|\delta_l^1 - \delta_l^2\right|}{2\tau + 1}\right)^2 + \max_{l=1,2,\ldots,T}\left(\frac{\left|\delta_l^1 - \delta_l^2\right|}{2\tau + 1}\right)^2\right)^{1/2} \quad (8)$$

Definition 16 The Taxicab distance between two terms, n_{ij} and ns_j, in m dimensional space can be defined as $T_i = \sum_{j=1}^{m} \left|n_{ij} - ns_j\right|$.

Definition 17 The Mahalanobis distance between two terms, n_{ij} and ns_j, in m dimensional space can be defined as:

$$\boldsymbol{M}_i = \sqrt{\left(\boldsymbol{n}_{ij} - \boldsymbol{ns}_j\right)\boldsymbol{C}^{-1}\left(\boldsymbol{n}_{ij} - \boldsymbol{ns}_j\right)} \quad (9)$$

where C^{-1} is the inverse covariance matrix.

Considering CODAS methodology Step 5, which establishes the calculation of a main and secondary distance of the negative ideal solution alternatives; definitions 11–15 are contemplated for the main distance calculation and, for this paper's proposal, definitions 16–17 are used for the secondary distance calculation. Afterward, to fulfill CODAS Step 7 a score will be defined from each evaluated alternative and a comparative analysis will be performed; thus, evaluating the reliability showed by the incorporation of Mahalanobis distance as the secondary distance measure for CODAS methodology.

5 CODAS-HFTLS-Mahalanobis Framework

Contributing to academic literature in horizontal collaboration, this chapter presents an analysis for hierarchizing horizontal collaboration factors using CODAS methodology with a new combinative distance measure equation. Centering data from an expert panel, whose experience in the academic field enhances horizontal collaboration factors for MCDM techniques. The expert panel is composed of five experts, from

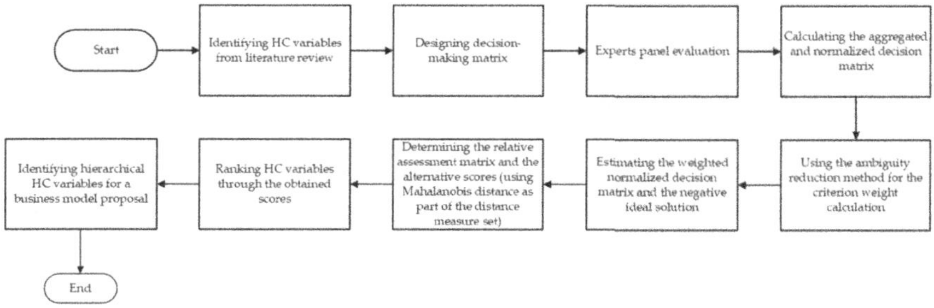

Fig. 1 Proposed MCDM flow diagram to identify HC variables' hierarchy

the disciplines of Industrial Engineering, Economics, Bioengineering, and Business Administration; with skills like innovation management, operations management, optimization, production planning, regional development, and business administration. Figure 1 conceptualizes the flow chart followed to reach this paper's main objective, considering the assigned linguistic terms, and the CODAS methodology utilizing Mahalanobis distance as part of the distance measure set for hierarchizing HC variables.

5.1 CODAS-HFLTS for Identifying Hierarchical HC Factors with Proposed Mahalanobis Distance

From the horizontal collaboration, factors were identified from a literature compendium and then, grouped within three business model components, presented in Sect. 3. From this point on, a decision-making matrix has been designed as of considering the 2018 OSLO Manual, selected for its approach of an innovation framework that can be applied to every type of economic activity [48] to identify the dimensions or capabilities needed to measure innovations activities. For the outcome in this chapter, just two out of the six components are taken into account (1) Internal dimension, all enterprise's capabilities acquired through time and experience,and, (2) External dimension, external factors that contribute or impact an enterprise's innovation. From these two components, five sets of criteria are determined for the MCDM decision matrix, presented in Table 5.

The decision-making matrix was presented to the expert panel, who provided an evaluation from the horizontal collaboration factors presented in Tables 1, 2, and 3. A linguistic terms scale was used to measure each factor, where meaning of Excellent (code S8, value 8), Very strong (code S7, value 7), Strong (code S6, value 6), Fine (code S5, value 5), Middle good (code S4, value 4), Unbiased (code S3, value 3), Medium insignificant (code S2, value 2), Insignificant (code S1, value 1) and, Null (code S0, value 0). Afterward, an aggregated and normalized matrix was obtained

Table 5 Innovation dimensions for the decision-making matrix [48]

Dimension/Types of capabilities	Elements
Internal dimension/Business capabilities	Business resources (BR) Management capabilities (MC) Workforce skills and human resource management (WS) Technological capabilities (TC)
External dimension/Measure of external factors	External business environment (EB)

from the expert panel evaluation, setting out the maximum value criteria (Step 1), as shown in Table 6.

Then, an ambiguity reduction is performed by developing a criterion weight calculation, as described in subsection 2.4.1. Reducing the availability of information gives certainty to researchers and decision-makers in a specific area [49]. The AR is being presented by [43] as a criterion weight calculation to reduce ambiguity. Table 7 displays the weighing values determined by the variable's frequency (w_j^{AK}).

To obtain the w_j^{AHP}, a pair-wise comparison matrix was developed through a 9-point scale, and then generating a standardized autovector to obtain wj, the normalized average value. Continuing with the consistency index by calculating $CI = (\lambda max - n_c)/(n_c - 1)$ and, the consistency ratio $CR = CI/RI$ which is accepted when is not greater than 10% of the random index (RI). Taking into account that nc = 5 and its equivalent RI = 1.12, the CI and RI obtained were 0.0959 and 8%, respectively demonstrating the consistency index of the AHP methodology. AHP weight assessment for each dimension is BR = 0.3419, MC = 0.2670, WS = 0.0835, TC = 0.1957, and EB = 0.1118. Table 8 displays the AR criteria assessment thus reducing the uncertainty between the information from previous researches and the expert panel evaluation.

The process following each criterion evaluation, is taken from the CODAS methodology, calculating the normalized decision matrix (Step 2), the weighted normalized decision matrix (Step 3), and the negative ideal solution (Step 4).

5.2 *Mahalanobis Distance*

Next, CODAS methodology Step 5 is obtained through proposing a series of distance sets for determining the alternative with greater distances. For the main distance measure sets, the authors present the collection of distance measures proposed by [47] for representing an expert panel hesitant preference in evaluating linguistics variables, as is the case developed in this paper.

Table 6 Aggregated decision matrix

Code	BR	MC	WS	TC	EB	Code	BR	MC	WS	TC	EB
C01	S8	S6	S7	S5	S6	E27	S8	S6	S7	S5	S6
C02	S7	S7	S7	S5	S7	E28	S7	S7	S6	S5	S8
C03	S7	S6	S6	S4	S7	E29	S8	S8	S7	S7	S6
C04	S6	S8	S7	S8	S6	E30	S5	S7	S8	S8	S6
C05	S7	S7	S8	S8	S8	E31	S6	S7	S8	S7	S5
C06	S6	S8	S8	S8	S5	E32	S7	S6	S7	S8	S5
C07	S8	S8	S6	S5	S6	E33	S7	S6	S7	S7	S6
C08	S8	S7	S8	S8	S8	E34	S8	S7	S7	S6	S8
C09	S8	S8	S7	S6	S7	E35	S8	S6	S7	S6	S7
C10	S7	S8	S7	S7	S6	E36	S4	S7	S7	S5	S6
C11	S8	S8	S7	S7	S8	G01	S4	S7	S8	S6	S8
C12	S7	S8	S7	S7	S7	G02	S4	S6	S6	S5	S7
C13	S8	S7	S8	S6	S5	G03	S7	S5	S5	S3	S8
C14	S7	S8	S7	S6	S7	G04	S5	S8	S6	S4	S5
C15	S8	S7	S8	S7	S8	G05	S6	S8	S8	S7	S7
C16	S8	S8	S6	S6	S8	G06	S7	S7	S8	S7	S7
C17	S5	S7	S8	S8	S6	G07	S6	S6	S7	S4	S5
C18	S6	S7	S8	S8	S6	G08	S5	S6	S8	S8	S8
C19	S7	S7	S7	S7	S6	G09	S6	S7	S8	S8	S6
C20	S7	S8	S7	S7	S7	G10	S6	S7	S8	S8	S8
C21	S8	S7	S7	S7	S6	G11	S6	S7	S8	S8	S6
C22	S5	S8	S8	S8	S6	G12	S8	S7	S6	S5	S6
C23	S8	S8	S7	S7	S7	G13	S7	S8	S6	S5	S6
C24	S7	S8	S8	S8	S6	G14	S6	S5	S8	S6	S7
C25	S8	S8	S8	S8	S6	G15	S6	S7	S8	S7	S6
C26	S6	S6	S7	S8	S6	G16	S6	S6	S7	S8	S7
C27	S6	S8	S7	S8	S6	G17	S6	S6	S5	S6	S6
C28	S8	S8	S7	S7	S6	G18	S8	S7	S5	S4	S7
C29	S7	S8	S7	S7	S8	G19	S8	S8	S7	S6	S8
C30	S8	S6	S5	S5	S6	G20	S7	S7	S8	S7	S7
E01	S6	S7	S8	S5	S8	G21	S7	S8	S6	S6	S7
E02	S6	S6	S5	S2	S6	G22	S8	S8	S7	S7	S6
E03	S8	S7	S7	S5	S6	G23	S7	S6	S8	S7	S7
E04	S8	S8	S7	S7	S8	G24	S6	S8	S6	S6	S6
E05	S7	S7	S7	S7	S7	G25	S7	S8	S6	S6	S6
E06	S8	S7	S8	S4	S5	G26	S6	S7	S7	S7	S7

(continued)

Table 6 (continued)

Code	BR	MC	WS	TC	EB	Code	BR	MC	WS	TC	EB
E07	S8	S8	S6	S6	S8	G27	S6	S7	S6	S6	S8
E08	S5	S8	S8	S8	S6	G28	S8	S7	S7	S6	S8
E09	S7	S8	S8	S7	S8	G29	S8	S7	S7	S6	S6
E10	S7	S7	S8	S6	S6	G30	S7	S7	S7	S6	S7
E11	S8	S8	S6	S5	S6	G31	S8	S7	S6	S6	S6
E12	S7	S8	S7	S6	S5	G32	S7	S8	S6	S5	S5
E13	S5	S7	S6	S6	S6	G33	S8	S6	S6	S8	S8
E14	S5	S8	S7	S7	S6	G34	S8	S8	S8	S7	S8
E15	S5	S7	S7	S7	S7	G35	S8	S8	S7	S7	S7
E16	S8	S6	S6	S5	S7	G36	S8	S7	S6	S6	S6
E17	S6	S6	S5	S5	S7	G37	S5	S8	S7	S6	S6
E18	S7	S7	S6	S6	S8	G38	S6	S7	S7	S7	S7
E19	S7	S7	S7	S6	S7	G39	S8	S8	S7	S7	S7
E20	S7	S8	S7	S5	S7	G40	S8	S7	S7	S7	S8
E21	S6	S8	S6	S4	S6	G41	S8	S8	S6	S7	S7
E22	S5	S8	S7	S7	S7	G42	S7	S7	S7	S7	S7
E23	S7	S8	S7	S7	S6	G43	S7	S7	S7	S7	S8
E24	S6	S8	S7	S7	S6	G44	S8	S6	S6	S4	S5
E25	S6	S6	S6	S6	S6	G45	S7	S8	S6	S7	S6
E26	S6	S7	S7	S8	S6	G46	S8	S7	S6	S6	S6

BR = Business resources, MC = Management capabilities, WS = Workforce skills & human resources management, TC = Technological capabilities, EB = External business management

Table 7 Acquired knowledge weight assessment

Business Model Criteria	BR	MC	WS	TC	EB	Total
Content	95	107	112	85	55	454
Structure	60	103	91	66	50	370
Governance	84	137	116	80	92	509
Score	239	347	319	231	197	1333
Weight (WAK)	**0.1793**	**0.2603**	**0.2393**	**0.1733**	**0.1478**	

Table 8 Ambiguity reduction criteria assessment

Criteria	BR	MC	WS	TC	EB
w_j^{AK}	0.1793	0.2603	0.2393	0.1733	0.1478
w_j^{AHP}	0.3419	0.2670	0.0835	0.1957	0.1118
w_j^{AR}	0.2606	0.2637	0.1614	0.1845	0.1298

- **Main distance measure:** D1: Hamming distance, D2: Hamming-Euclidean distance, D3: Hamming-Hausdorff distance, D4: Hybrid Hamming-Hausdorff distance, and D5: Hybrid Euclidean Hamming-Hausdorff distance.

  Then, for the secondary distance measure set, the authors propose Mahalanobis distance and its generalizations against Taxicab distance measure, on the assumption that Mahalanobis distance provides a higher coefficient against Taxicab distance. The Mahalanobis distance is characterized by presenting the same distance between neighbors within a large multivariate data, e.g., stretching the sphere from a set of points correcting the scales of the variables and their correlation [50, 51, 52].
- **Secondary distance measure:** T1: Taxicab distance, M1: Mahalanobis (Cov($\overline{r}_{ij}$),$\overline{r}_{ij}$), M2: Mahalanobis (Cov($\overline{r}_{ij}$),ns), and M3: Mahalanobis (Cov(ns),ns), all three M1,2,3 distances from Definition 17.

Continuing with CODAS methodology, these distance combination sets were used to construct the relative evaluation matrix (Step 6), from which the preference score was obtained for each evaluated alternative (Step 7) and; finally, displaying the hierarchize HC variables (Step 8). Table 8 shows the top 20 hierarchized horizontal collaboration alternatives from each distance combination.

From Table 8, a concordance analysis was performed for each Mahalanobis generalization, contrasting against the Taxicab distance results, used traditionally in CODAS methodology. Table 9 displays the aforementioned where it can be observed that M_2 results maintain a higher proportion in concordance compared to T_1 results.

5.3 *Sensitivity Analysis*

For the sensitivity analysis, the preference scores previously calculated were used. The Cronbach's alpha coefficient obtained where 0.9542 from T_1 distance; 0.9528 from M_1 distance; 0.9548 from M_2 distance; and, 0.9541 from M_3 distance, all used as the secondary distance measure. The previous values show the internal high consistency from the obtained calculations; however, it is important to observe that the M_2 distance displays a higher coefficient against the other distances for the secondary distance measure.

Similar results are presented by analyzing the correlation matrix for each main and secondary distance combination set, where by generalizing them, an average value of correlations is obtained as follows 0.884 from T_1 distance; 0.883 from M_1 distance; 0.890 from M_2 distance; and, 0.887 from M_3 distance. These results are considered high in every case, nevertheless for the M_2 distance is slightly higher than the rest.

Figure 2 displays the comparison regarding which main distance measure provides better results against the secondary distance measure sets. The data is presented in Table 9 from a graphic point of view. From this graphic, it can be observed a 0.94 from D_1 distance; 0.91 from D_4 distance; 0.73 from D_2 distance; 0.52 from D_5

Table 9 Top 20 alternatives obtained using T1, M1, M2 and M3 distances

	T_1					M_1					M_2					M_3				
Rank	D_1	D_2	D_3	D_4	D_5	D_1	D_2	D_3	D_4	D_5	D_1	D_2	D_3	D_4	D_5	D_1	D_2	D_3	D_4	D_5
1	G34	C25	C26	C25	C25	G34	C25	C18	C25	C25	G34	C25	C17	C25	C25	G34	C25	C24	C25	C25
2	C25	C08	G16	C08	C08	C25	C08	G09	C08	C08	C25	C08	E30	C08	C08	C25	C08	C25	C08	C08
3	C08	G34	E32	C24	C24	C08	G34	G11	C24	C24	C08	G34	C26	C24	C24	C08	G34	C18	C24	C24
4	C11	C11	G08	G34	G33	C11	C11	C24	G34	C05	C11	C11	E26	C05	C05	C11	C11	G09	G34	C05
5	E04	E04	C17	C05	C05	E04	E04	E26	C05	C04	E04	E04	G16	G34	C06	E04	E04	G11	C05	C06
6	E09	C23	E30	C11	C06	E09	C23	C04	C11	C27	E09	C23	C18	C11	G33	E09	C23	C05	C11	C04
7	C15	G35	E26	E04	C04	C15	G35	C27	E04	C06	C15	G35	G09	E04	C04	C15	G35	C08	E04	C27
8	C23	G39	C18	C15	C27	C23	G39	C05	C23	G33	C23	G39	G11	G10	C27	C23	G39	E26	C23	G33
9	G35	C24	G09	G10	C22	G35	C24	C17	G35	G10	G35	C24	C22	C15	C22	G35	C24	C04	G35	G10
10	G39	C28	G11	C23	E08	G39	C28	E30	G39	C22	G39	C28	E08	C23	E08	G39	C28	C27	G39	C22
11	C24	E29	C22	G35	G10	C24	E29	G10	C15	E08	C24	E29	G08	G35	G10	C24	E29	G10	C15	E08
12	C05	G22	E08	G39	E32	C05	G22	C25	G10	C18	C05	G22	C04	G39	C18	C05	G22	C17	G10	C18
13	G19	G41	C04	G33	C18	G19	G41	C22	G19	G09	G19	G41	C27	G19	G09	G19	G41	E30	G19	G09
14	C28	C15	C27	G19	G09	C28	C15	E08	G33	G11	C28	C15	E32	G33	G11	C28	C15	C22	G33	G11
15	E29	G33	C06	C06	G11	E29	G33	C08	C28	E26	E29	G33	C06	C06	E26	E29	G33	E08	C28	E26
16	G22	G19	G33	C28	E26	G22	G19	C06	E29	C17	G22	G19	G10	C28	E32	G22	G19	C06	E29	C17
17	C29	C05	G10	E29	C17	C29	C05	G16	G22	E30	C29	C05	C24	E29	C17	C29	C05	G16	G22	E30
18	G40	G40	C05	G22	E30	G40	G40	C26	C06	E32	G40	G40	C05	G22	E30	G40	G40	C26	G40	E32
19	G41	E09	C24	C04	G08	G41	E09	G08	G40	G16	G41	E09	C25	C04	G16	G41	E09	G08	C06	G16
20	C09	C16	C08	C27	G16	C09	C09	E32	C04	C26	C09	C16	G33	C27	C26	C09	C09	E32	C04	C26

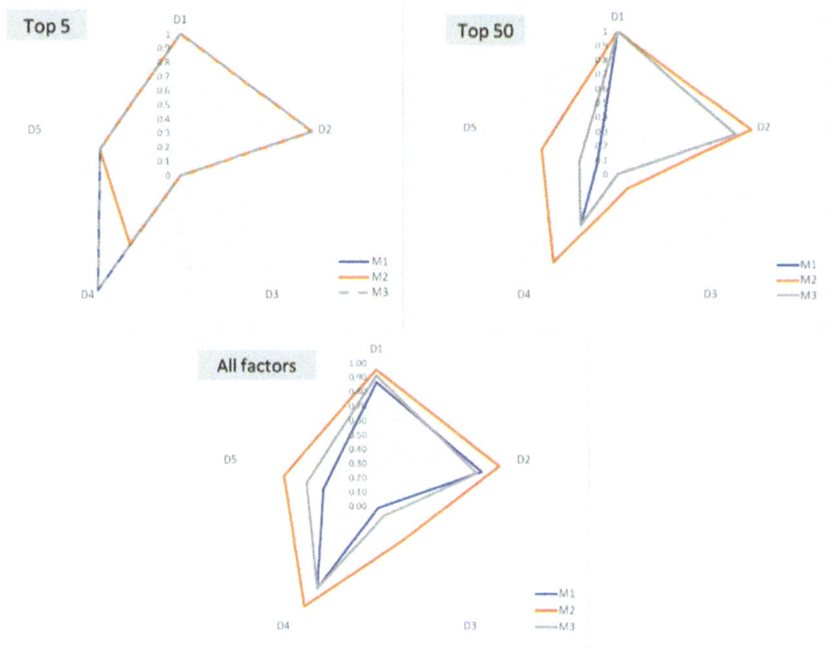

Fig. 2 Concordance analysis from Taxicab distance versus Mahalanobis distances

distance; and, 0.06 from D_3 distance. The above sustains the proposal to use the Hamming and Mahalanobis (Cov($\overline{r}_{ij}$),ns) distances, as the primary and secondary distance measures, respectively, for the CODAS methodology Step 5.

6 COHRV Business Model for a Value Creation Network

The previous section describes the process from which HC variables have been hierarchize; analyzing the best distance measure set, which resulted in Hamming-Mahalanobis (Cov($\overline{r}_{ij}$),ns) distances, as the primary and secondary distance measures, respectively, for CODAS Step 5. From this point forward, the HC variables now listed through a preference score can be identified through the obtained rank.

From the authors' preference, the Top 30 HC factors are identified by ranking them in an ascending form. Creating a model for enterprises to establish responsiveness and a flexible supply chain towards a sustainable manufacturing 4.0 environment. Table 10 displays the Top 30 HC factors obtained from the Hamming-Mahalanobis CODAS-HFTLS methodology.

Enterprises participating in a joint venture can add experience, flexibility and teamwork; as well as, complementing each other capabilities [29, 36]. From the

Table 10 Taxicab distance versus Mahalanobis distance concordance analysis

(T_1)	M1			M2			M3			
	Top 5	Top 50	All data	Top 5	Top 50	All data	Top 5	Top 50	All data	Average
D_1	1.00	0.86	0.87	1.00	1.00	0.96	1.00	0.88	0.91	0.94
D_2	1.00	0.92	0.78	1.00	0.98	0.91	1.00	0.84	0.74	0.91
D_3	0.00	0	0.02	0.00	0.12	0.30	0.00	0.00	0.09	0.06
D_4	1.00	0.48	0.71	0.60	0.74	0.87	1.00	0.48	0.71	0.73
D_5	0.60	0.26	0.39	0.60	0.62	0.69	0.60	0.44	0.52	0.52
Average	**0.72**	**0.50**	**0.55**	**0.64**	**0.69**	**0.75**	**0.72**	**0.53**	**0.59**	

above list, the horizontal collaboration factors of higher priority are grouped within three components (1) Concept, 13 variables; (2) Structure, 5 variables; and, (3) Governance, 12 variables (Table 11).

6.1 Content Component

The Content component groups the *what* activities to implement a horizontal collaboration model. From the MCDM analysis, thirteen horizontal collaborations have been hierarchized, being:

1. **Firm performance (score −2.1430)**, measuring enterprises' performance throughout the horizontal collaboration to follow improvement or growth, in innovation, financial and operational level, cost, quality, and delivery improvements [7, 35].
2. **Production flexibility (score −0.2.1288)**, enterprises must consider volume or capacity flexibility, as production changes through the alliance project.
3. **Trust (score −1.9493)**, a condition required for horizontal collaboration to occur; enhancing trust by sharing information, efforts, and resources throughout a focused partner selection with similarities, so the connection can be built [53].
4. **Operational performance (score −1.1702)**, combining and adjusting operations may lead to reducing supply chain errors, costs, and time.
5. **Goal alignment (score −1.6551)**, a mutual definition of goals must be taken into account; thus, providing motivation and recognition of fairness [33].
6. **Research & Development (score −1.5480)**, defining specific market demand and knowledge, leading to innovative products or services within the HC.
7. **Process performance (score −1.5334)**, integrating processes and capacities seeking efficiency, and information sharing for supply chain integration. Training and knowledge transfer may occur for this to gain the proper structure.

Table 11 Top 30 horizontal collaboration factors through Hamming-Mahalanobis CODAS-HFTLS methodology

	Hamming-Mahalanobis Codas HFLTS		
HC variables	Code	Score	Ranking
Confidentiality	G34	−2.315826554	1
Firm performance	C25	−2.143032496	2
Production flexibillity	C08	−2.128808161	3
Trust	C11	−1.949310563	4
Process performance	E04	−1.949310563	5
Quality	E09	−1.721317641	6
Operational performance	C15	−1.710244678	7
Goal alignment	C23	−1.655163687	8
Business strategy	G35	−1.655163687	9
Absorption capacity	G39	−1.655163687	10
Research and Development	C24	−1.548059985	11
Process performance	C05	−1.533405156	12
Trust	G19	−1.522122279	13
Commercialization phase	C28	−1.34714842	14
Communication	E29	−1.34714842	15
Leadership	G22	−1.34714842	16
Value creation	C29	−1.344040601	17
Goal alignment	G40	−1.334679522	18
Business model identification	G41	−1.262313667	19
Focus strategy in limited resources	C09	−1.222276552	20
Negotiation performance	C16	−1.134999034	21
Costs	E07	−1.134999034	22
Performance metrics	C12	−1.041265189	23
Waste reduction	C20	−1.041265189	24
Quality	G10	−0.916119543	25
Market strategy	E34	−0.898332373	26
Commitment	G28	−0.898332373	27
Process performance	G05	−0.80055555	28
Problem solving and support	G06	−0.79440757	29
Cooperation	G20	−0.79440757	30

8. **Commercialization phase (score −1.3471)**, planning ahead for this phase will bring trust and guidance for the innovative product and service designed. Enterprises must take into account launching time, market testing, and after-sales support, among others [39].

9. **Value creation (score −1.3440)**, enhancing value creation from the alliance, cultivating a collaborative culture where enterprises acknowledge the relationship and each firm's responsibility within the HC.
10. **Focus strategy on limited resources (score −1.2222)**, bringing forward each enterprise's limited resources; e.g., production quality, marketing sales, and design capability, among others. From this point forward, partners can build a specific content for a horizontal collaboration value creation network.
11. **Negotiation performance (score −1.1349)**, joint definition of rules, contracts, legal protection where enterprises align strategic agreements involving operational, financial, and legal areas.
12. **Performance metrics (score −1.0412)**, identifying desirable key outputs, such as waste, inventory, costs reduction; increased quality in delivery times, production processes, and bringing improvements to the new value creation network [22].
13. **Waste reduction (score −1.0412)**, emphasizing market requirements with the purpose of obtaining optimization in production lines, delivery time, logistics; gaining reduction in waste, $CO2$, inventory, and production turnover.

6.2 Structure Component

The Structure component groups the activities for the how to link enterprises in the collaboration. From the analysis, the Structure component listed 5 factors, described below:

1. **Process performance (score −1.9493)**, specifying which process will be combined, redesigned for reducing instability, and improving efficiency and flexibility within the new value creation network.
2. **Quality (score −1.7213)**, the link between the performance metrics defined in the horizontal collaboration project; e.g., waste elimination, customer complaints, continuous product and service monitoring, and delivery time, among others [26].
3. **Communication (score −1.3471)**, implementing IT infrastructure, enhancing multidisciplinary team information and knowledge exchange to strengthening enterprises communication within the horizontal collaboration project.
4. **Costs (score −1.1349)**, achieving reduction, economies of scale, distribution, and logistics costs; considering an overall supply chain costs element through implementing a horizontal collaboration value creation network.
5. **Market strategy (score −0.8983)**, creating value inside the market by developing new technology, products, or services that establish a leading position, focusing horizontal collaboration enterprises in a win–win situation [9].

6.3 Governance Component

The last component from the business model perspective, includes activities that focus on who has to take action within the horizontal collaboration. From the analysis, the Governance component contains 12 factors, such as:

1. **Confidentiality (score −2.3158)**, enterprises multidisciplinary team must take into account confidentiality as a core value within the new network, for horizontal collaboration to work effectively.
2. **Business strategy (score −1.6551)**, enterprises can leverage liabilities by developing a business strategy from a multidisciplinary team that works toward gaining market opportunities.
3. **Absorption capacity (score −1.6551)**, enterprises must partner taking into account the capacity of decision-making from the management level. Which firm, owners, have a promotion and decision-making towards inter-organizational ventures, investing in R&D, improvements, multidisciplinary teams for creating innovative products and services [54].
4. **Trust (score −1.5221)**, a key element that will evolve during the horizontal collaboration relationship, gaining strength by relying in each partner, according to goal alignment.
5. **Leadership (score −1.3471)**, there's an enterprise with a higher leadership standard, who must work for the greater good of the horizontal collaboration project, focusing strategies, multidisciplinary teams, and resources towards the aligned goals and benefits.
6. **Goal alignment (score −1.3346)**, defined as a content component, goal alignment must be down-scaled to the multidisciplinary team so they work focused in each goal assigned [28].
7. **Business model identification (score −1.2623)**, as a sustainable manufacturing 4.0 presents a disruption in business model, enterprises working jointly must determine who will enact in the new supply chain, how will they link each department, process, and innovation towards aiming for the new value network.
8. **Quality (score −0.9161)**, multidisciplinary teams must fulfill the quality component, being certified, sharing, and training the rest of the team for search quality in every aspect of the horizontal collaboration project.
9. **Commitment (score −0.8983)**, a level of commitment among enterprises must be taken into account and build upon the horizontal collaboration, for a supply chain sustainability and a sharing of information, innovation, process, and operational performance [35].
10. **Process performance (score −0.8005)**, a multidisciplinary team must be created thinking about the management of the new, improve or share processes within the horizontal collaboration.
11. **Problem solving and support (score −0.7944)** , enterprises must partner up taking into account a team capable of supporting and solving the horizontal collaboration new problems ahead; e.g., IP legal work, supply chain network, management paperwork, among others.

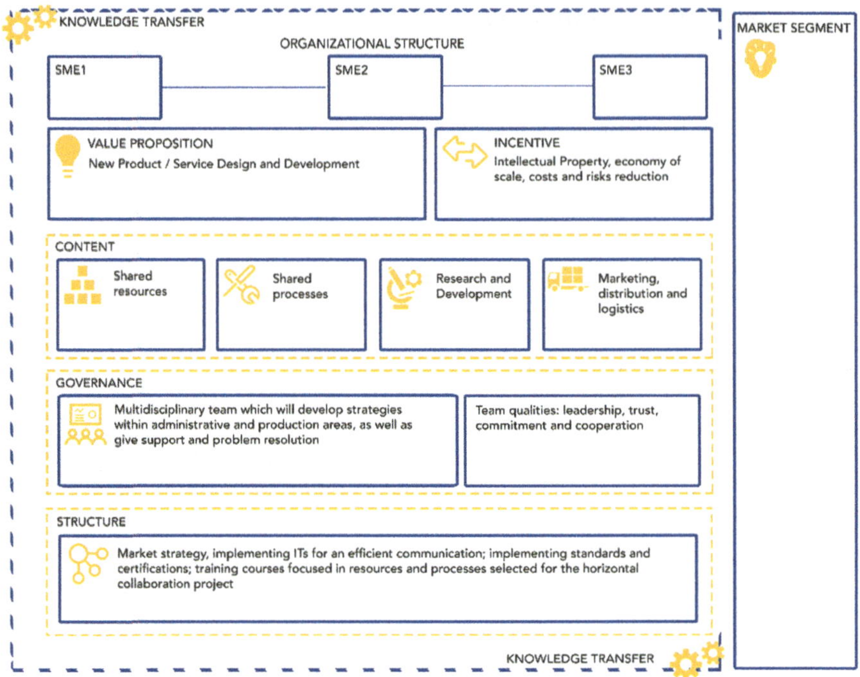

Fig. 4 COHRV Model for the deployment of horizontal collaboration strategies

12. **Cooperation (score −0.7944)**, enterprise size and organizational culture similarities gain strength allowing a mutual coordination toward knowledge transfer and innovation [54].

The authors' horizontal collaboration business model (COHRV Model) contributes toward a disruptive business model presenting horizontal collaboration as a framework for enterprises to develop strategies for innovation and new product development. Figure 4 shows the four blocks of business model COHRV where each horizontal collaboration factor identified as of higher priority are considered for enterprises to deploy strategies for joint projects and develop a value creation network from a business model that contemplates two or more enterprises.

7 Conclusions

As the supply chain gains globalization or suffers from a healthcare pandemic, enterprises must focus on strategies for the challenge it represents to implement new strategies, technologies, or partnerships. From the lack of knowledge to the investment cost, enterprises must find alternatives to upgrade themselves inside their supply chain network. Horizontal collaboration is been seen as the new vertical integration, where competitors work jointly towards a common goal.

Literature provides two perspectives (a) theoretical views, where information is gathered and displayed from a qualitative or meta-analysis; or (b) empirical views, in which data is collected from case studies in certain sectors, i.e., logistics service providers, or by type of enterprises regarding size, geographic area or supply chain. Researchers display several variables for its field use. Although, they tend to be distinct between each research; presenting a few similar ones like information sharing, decision synchronization, trust, and partner selection. Nonetheless, it lacks a model or methodology for enterprises, to plan and perform a horizontal collaboration in short to long-term relationship.

This chapter presents a model responding to the what, how and who has to contribute to the horizontal collaboration, by presenting a Top 30 list clustered within three business model components (1) Content, (2) Structure and, (3) Governance. This is achieved by gathering items from a literature compendium and, evaluating them through an MCDM method. This is accomplished by proposing a horizontal collaboration hierarchy through the Hamming-Mahalanobis ($\text{Cov}(\overline{r}_{ij})$, ns) distance from a CODAS-HFLTS methodology. For this to be reached, an analysis was developed through each set of distance measure described and followed in CODAS-HFLTS methodology. Mahalanobis distance ($\text{Cov}(\overline{r}_{ij})$, ns) was selected as the secondary distance measure set, resulting in a higher coefficient against the other distances. Mahalanobis distance takes into account the correlation between variables [38, 55] that can be identified and analyzed relative to a base group [56]; hence, highlighting a solution for an MCDM through a combinative distance-based assessment.

The Top 30 list, facilitates the planning and development of a horizontal collaboration, giving reassurance and trust in the project ahead. Laying foundations for the horizontal collaboration as one of Industry 4.0 elements, through a disruptive business model for sustainable manufacturing 4.0 to be performed. Therefore, creating an environment where innovation can be achieved by making accessible new products and services to customers, gaining permanence or position within the supply chain. This chapter contributes to the sustainable manufacturing 4.0 literature by proposing a disruptive horizontal collaboration business model in a twofold: first, conceptualizing an initial list of horizontal collaboration factors, clustering them in three business model components, which helps structure the what, how and who can contribute within an alliance. Thereby, facilitating enterprises the developing of a horizontal collaboration from a business model perspective, easing up trust between the participating enterprises. Second, the horizontal collaboration factors list is put through an MCDM technique proposed by the authors, by implementing Mahalanobis distance as the secondary distance measure set from a CODAS-HFLTS method, enhancing the importance in using Mahalanobis as a distance that's capable of measuring the distance between several variables by contemplating its correlation; and by this, obtaining a Top 30 list.

References

1. Barratt M (2004) Understanding the meaning of collaboration in the supply chain. Supply Chain Manag Int J 9(1):30–42. https://doi.org/10.1108/13598540410517566
2. Simatupang TM, Sridharan R (2002) The Collaborative Supply Chain. Int J Logistics Manag 13(1):15–30. https://doi.org/10.1108/09574090210806333
3. Cruijssen F, Cools M, Dullaert W (2007) Horizontal cooperation in logistics: opportunities and impediments. Transp Res Part E Logistics Transp Rev 43(2):129–142. https://doi.org/10.1016/j.tre.2005.09.007
4. Derrouiche R, Gamoura S, Damand D, Bouguessas H (2018) Towards a collaborative supply chain balanced score card framework to analyse collaborative value-added. In: IEEE international conference on industrial engineering and engineering management, 2017 Dec, pp 427–431. https://doi.org/10.1109/IEEM.2017.8289926
5. Whitehead K, Zacharia Z, Prater E (2019) Investigating the role of knowledge transfer in supply chain collaboration. Int J Logist Manag 30(1):284–302. https://doi.org/10.1108/IJLM-07-2017-0184
6. Osterwalder A, Pigneur Y (2010) Business model generation. John Wiley & Sons Inc., Hoboken, New Jersey
7. Yılmaz H, Çemberci M, Uca N (2016) The role of collaborative advantage for analyzing the effect of supply chain collaboration on firm performance. Int J Commer Finance 2(1):157–168
8. Agostini L, Nosella A (2019) Inter-organizational relationships involving SMEs: a bibliographic investigation into the state of the art. Long Range Plan 52(1):1–31. https://doi.org/10.1016/j.lrp.2017.12.003
9. Ahn JM, Kim D, Moon S (2017) Determinants of innovation collaboration selection: a comparative analysis of Korea and Germany. Innovation 19(2):125–145. https://doi.org/10.1080/14479338.2016.1241152
10. Rendón A, Morales A et al (2010) Alianzas y adquisiciones. Estrategias de competitividad de las grandes empresas de capital privado nacional en México, RIICO, 978-970-27-2032-4. https://www.riico.net/index.php/riico/article/view/738
11. Naesens K, Pintelon L, Taillieu T (2007) A framework for implementing and sustaining trust in horizontal partnerships. Supply Chain Forum Int J 8(1):32–44. https://doi.org/10.1080/16258312.2007.11517174
12. Amit RH, Zott C (2012) Creating value through business model innovation. MIT Sloan Manag Rev 53(3):40–50. https://doi.org/10.2139/ssrn.1701660
13. Baden-Fuller C, Morgan MS (2010) Business models as models. Long Range Plan 43(2–3):156–171. https://doi.org/10.1016/j.lrp.2010.02.005
14. Chesbrough H (2010) Business model innovation: opportunities and barriers. Long Range Plan 43(2–3):354–363. https://doi.org/10.1016/j.lrp.2009.07.010
15. Saur-Amaral I, Soares RR, Proença JF (2018) Business model innovation: towards a conceptual framework. Tourism Manag Stud 14(1):80–93. https://doi.org/10.18089/tms.2018.14107
16. Wirtz BW, Pistoia A, Ullrich S, Göttel V (2016) Business models: origin, development and future research perspectives. Long Range Plan 49(1):36–54. https://doi.org/10.1016/j.lrp.2015.04.001
17. Fischer M, Imgrund F, Janiesch C, Winkelmann A (2020) Strategy archetypes for digital transformation: defining meta objectives using business process management. Inf Manag 57(5):103262. https://doi.org/10.1016/j.im.2019.103262
18. Flanagan DJ, Lepisto DA, Ofstein LF (2018) Coopetition among nascent craft breweries: a value chain analysis. J Small Bus Enterp Dev 25(1):2–16. https://doi.org/10.1108/JSBED-05-2017-0173
19. Franco VB, Arboleda CDP, Cantillo V, Macea LF, Ríos DGR (2017) A collaborative supply chain model for non-for-profit networks based on cooperative game theory. Int J Logistics Syst Manage 26(4):475. https://doi.org/10.1504/IJLSM.2017.082614

20. Leitner R, Meizer F, Prochazka M, Sihn W (2011) Structural concepts for horizontal cooperation to increase efficiency in logistics. CIRP J Manuf Sci Technol 4(3):332–337. https://doi.org/10.1016/j.cirpj.2011.01.009
21. Pomponi F, Fratocchi L, Rossi Tafuri S (2015) Trust development and horizontal collaboration in logistics: a theory based evolutionary framework. Supply Chain Manag Int J 20(1):83–97. https://doi.org/10.1108/SCM-02-2014-0078
22. Sanchez Rodrigues V, Harris I, Mason R (2015) Horizontal logistics collaboration for enhanced supply chain performance: an international retail perspective. Supply Chain Manag Int J 20(6):631–647. https://doi.org/10.1108/SCM-06-2015-0218
23. de Sousa L, Jabbour AB, Alves Filho AG, Noronha Viana AB, Chiappetta Jabbour CJ (2011) Factors affecting the adoption of supply chain management practices: evidence from the Brazilian electro-electronic sector. IIMB Manag Rev 23(4):208–222. https://doi.org/10.1016/j.iimb.2011.09.001
24. Björnfot A, Torjussen L (2012) Extent and effect of horizontal supply chain collaboration among construction SME. J Eng Project Prod Manag 2(1):47–55
25. Aggarwal S, Srivastava MK (2016) Towards a grounded view of collaboration in Indian agri-food supply chains. Br Food J 118(5):1085–1106. https://doi.org/10.1108/BFJ-08-2015-0274
26. Levi-Bliech M, Naveh G, Pliskin N, Fink L (2018) Mobile technology and business process performance: the mediating role of collaborative supply-chain capabilities. Inf Syst Manag 35(4):308–329. https://doi.org/10.1080/10580530.2018.1503803
27. Amer LE, Eltawil AB (2015) Analysis of quantitative models of horizontal collaboration in supply chain network design: Towards "green collaborative" strategies. In: International Conference on Industrial Engineering and Operations Management (IEOM), pp. 1–10. https://doi.org/10.1109/IEOM.2015.7093759
28. Pomponi F, Fratocchi L, Tafuri SR, Palumbo M (2013) Horizontal collaboration in logistics : a comprehensive framework. Res Logistics Prod 3:243–254
29. Prakash A, Deshmukh SG (2010) Horizontal collaboration in flexible supply chains: a simulation study. J Stud Manuf 1(January):54–58. http://www.hypersciences.org/JSM/Iss.1-2010/JSM-6-1-2010.pdf
30. Shafique MN, Khurshid MM, Rahman H, Khanna A, Gupta D, Rodrigues JJPC (2019) The role of wearable technologies in supply chain collaboration: a case of pharmaceutical industry. IEEE Access 7:49014–49026. https://doi.org/10.1109/ACCESS.2019.2909400
31. van Lier T, Caris A, Macharis C (2016) Sustainability SI: bundling of outbound freight flows: analyzing the potential of internal horizontal collaboration to improve sustainability. Netw Spat Econ 16(1):277–302. https://doi.org/10.1007/s11067-014-9226-x
32. He Z, Ouyang J (2011) SME cluster development strategy of supply chain. In: 2nd International conference on artificial intelligence, management science and electronic commerce (AIMSEC), pp 4770–4772. https://doi.org/10.1109/AIMSEC.2011.6010161
33. Daudi M, Hauge JB, Thoben K-D (2016) Behavioral factors influencing partner trust in logistics collaboration: a review. Logist Res 9(1):19. https://doi.org/10.1007/s12159-016-0146-7
34. Hudnurkar M, Jakhar S, Rathod U (2014) Factors affecting collaboration in supply chain: a literature review. Proc Soc Behav Sci 133:189–202. https://doi.org/10.1016/j.sbspro.2014.04.184
35. Shin N, Park SH, Park S (2019) Partnership-based supply chain collaboration: impact on commitment, innovation, and firm performance. Sustainability (Switzerland) 11(2). https://doi.org/10.3390/su11020449
36. Sitadewi D, Okdinawati L (2018) The conceptual framework of horizontal collaborative transportation management in Indonesian trucking industry, pp 42–58
37. Simatupang TM, Sridharan R (2008) Design for supply chain collaboration. Bus Process Manag J 14(3):401–418. https://doi.org/10.1108/14637150810876698
38. Martin N, Verdonck L, Caris A, Depaire B (2018) Horizontal collaboration in logistics: decision framework and typology. Oper Manag Res 11(1–2):32–50. https://doi.org/10.1007/s12063-018-0131-1

39. Wu Q, Luo X, Slotegraaf RJ, Aspara J (2015) Sleeping with competitors: the impact of NPD phases on stock market reactions to horizontal collaboration. J Acad Mark Sci 43(4):490–511. https://doi.org/10.1007/s11747-014-0396-3
40. Keshavarz Ghorabaee M, Zavadskas EK, Turskis Z, Antucheviciene J (2016) A new combinative distance-based assessment (CODAS) method for multi- criteria decision-making. Econom Comput Econom Cybernet Stud Res 50(3):25–44
41. Zhang X, Xu Z (2014) Extension of TOPSIS to multiple criteria decision making with pythagorean fuzzy sets. Int J Intell Syst 29(12):1061–1078. https://doi.org/10.1002/int.21676
42. Villa Silva AJ, Pérez Dominguez LA, Martínez Gómez E, Alvarado-Iniesta A, Pérez Olguín IJC (2019) Dimensional analysis under pythagorean fuzzy approach for supplier selection. Symmetry 11(3). https://doi.org/10.3390/sym11030336
43. Sansabas-Villalpando V, Pérez-Olguín IJC, Pérez-Domínguez LA, Rodríguez-Picón LA, Mendez-González LC (2019) CODAS HFLTS method to appraise organizational culture of innovation and complex technological changes environments. Sustainability 11(24):7045. https://doi.org/10.3390/su11247045
44. Badi I, Ballem M, Shetwan A (2018) Site selection of desalination plant in Libya by using combinative distance-based assessment (CODAS) method. Int J Qual Res 12(3):609–624. https://doi.org/10.18421/IJQR12.03-04
45. Aguarón-Joven J, Escobar-Urmeneta MT, García-Alcaraz JL, Moreno-Jiménez JM, Vega-Bonilla A (2015) A new synthesis procedure for TOPSIS based on AHP. DYNA. https://doi.org/10.15446/dyna.v82n191.51140
46. Rodriguez RM, Martinez L, Herrera F (2012) Hesitant fuzzy linguistic term sets for decision making. IEEE Trans Fuzzy Syst 20(1):109–119. https://doi.org/10.1109/TFUZZ.2011.2170076
47. Liao H, Xu Z, Zeng X (2014) Distance and similarity measures for hesitant fuzzy linguistic term sets and their application in multi-criteria decision making. Inf Sci. https://doi.org/10.1016/j.ins.2014.02.125
48. OECD (2018) Oslo Manual 2018: Guidelines for collecting, reporting and using data on innovation. In: The measurement of scientific, technological and innovation activities, 4th edn. https://doi.org/10.1787/9789264304604-en
49. Brun E (2008) Ambiguity reduction in new product development projects. Int J Innov Manag 12(4):573–596. https://doi.org/10.1142/S1363919608002126
50. Ali H, Syed Yahaya SS (2013) On robust mahalanobis distance issued from minimum vector variance. Far East J Math Sci 74(2):249–268
51. Dokas P, Ertoz L, Kumar V, Lazarevic A, Srivastava J, Tan P-N (2002) Data mining for network intrusion detection. In: National science foundation workshop on next generation data mining 38(7):21–30. https://doi.org/10.1111/j.1540-5915.2001.tb00975.x
52. Li SZ, Jain A (eds) (2009) Mahalanobis distance. In: Encyclopedia of biometrics. Springer, Boston, MA. https://doi.org/10.1007/978-0-387-73003-5_865
53. Gao H, Yang J, Yin H, Ma Z (2017) The impact of partner similarity on alliance management capability, stability and performance. Int J Phys Distrib Logist Manag 47(9):906–926. https://doi.org/10.1108/IJPDLM-09-2016-0263
54. Pérez CD, Ozuna AA, Arriaga AA (2011) Clustering and innovation capabilities in the mexican software industry. EMJ Eng Manage J 23(4):47–56. https://doi.org/10.1080/10429247.2011.11431919
55. Escobedo Portillo MT, Plata JAS (2008) Mahalanobis y las aplicaciones de su distancia estadistica. Estadística 5(27):13–20
56. Durán Hernández F, Rodríguez Medina MA, Rodríguez Morachis M, Pérez Olguín IJC (2016) Aplicación del Sistema Mahalanobis-Taguchi (MTS), para Determinar Variables Significantes que Ayuden a Definir Perfiles de Éxito en Alumnos Egresados de la UTCJ. La Investigación Como Herramienta Del Desarrollo, pp 183–189
57. Jaharuddin NS, Mansor ZD, Yaakob S (2016) Assessing the supply chain intelligence practices of small medium enterprises in Malaysia. Proc Econ Finance 35:515–521. https://doi.org/10.1016/S2212-5671(16)00064-2

58. Kohli AS, Jensen JB (2010) Assessing effectiveness of supply chain collaboration: an empirical study. Supply Chain Forum Int J 11(2):2–16. https://doi.org/10.1080/16258312.2010.11517228
59. Mojarro M, Pérez RJ, García JL (2007) Comparación de la técnica TOPSIS tradicional y la técnica TOPSIS con la distancia de Mahalanobis. Caso de estudio. Actitudestrascender
60. Pfeifer S, Peterka SO, Stanić M (2017) Business models of micro businesses: empirical evidence from creative industries. J Contemp Manage 22:1–19
61. Salam MA (2017) The mediating role of supply chain collaboration on the relationship between technology, trust and operational performance. Benchmark Int J 24(2):298–317. https://doi.org/10.1108/BIJ-07-2015-0075
62. Siong Choy C, Kuan Yew W, Lin B (2006) Criteria for measuring KM performance outcomes in organisations. Ind Manag Data Syst 106(7):917–936. https://doi.org/10.1108/02635570610688850
63. Wind Y, Saaty TL (1980) Marketing applications of the analytic hierarchy process. Manage Sci 26(7):641–658. https://doi.org/10.1287/mnsc.26.7.641
64. Xue X, Wang S, Lu B (2016) Manufacturing service composition method based on networked collaboration mode. J Netw Comput Appl 59:28–38. https://doi.org/10.1016/j.jnca.2015.05.003

Assessment of Industry 4.0 Adoption for Sustainability in Small and Medium Enterprises: A Fermatean Approach

Mahyar Kamali Saraji and Dalia Streimikiene

1 Introduction

The manufacturing industry's complexity and demand have constantly expanded recently. Increasing worldwide competitiveness, market instability, the desire for highly customized products, and reduced product life cycles are formidable obstacles for businesses; Therefore, it is believed that current approaches for value generation no longer meet the rising demands for cost-effectiveness, adaptability, sustainability, and stability [1, 2]. As a result, against this backdrop, trends and new buzzwords such as internet of services (IoS), digitalization, internet of things (IoT), and cyber-physical systems (CPS) have become increasingly remarkable, motivating Germany to launch Industry 4.0 as part of its high-tech strategy in 2011 [3]. Industry 4.0 is a transition in the manufacturing industry, and it introduces an entirely new viewpoint on how production might work with emerging technologies to achieve maximum output with minimal resource use [4]. Furthermore [5], concluded that Industry 4.0 could be defined by basic concepts: decentralization, interoperability, real-time capacity, modularity, virtualization, and service orientation. Decentralization is the continuous interchange of information that enables cyber-physical systems to make choices autonomously in real-time [6]. Interoperability strives to facilitate system-to-system communication, allowing users to embrace open standards [7]. The real-time capacity entails instantaneous data collection and swift and agile decision-making facilitation [8]. Modularity is defined by the adaptability of the whole manufacturing process, which permits the reorganization of production lines by linking and decoupling modules [9]. Virtualization facilitates the production of cloud-based virtual versions of physical systems, enabling simulation operations in real-time [6]. Service

M. Kamali Saraji · D. Streimikiene (✉)
Kaunas Faculty, Vilnius University, Muitines 8, 44280 Kaunas, Lithuania
e-mail: Dalia.Streimikiene@knf.vu.lt

M. Kamali Saraji
e-mail: Mahyar.Kamali@knf.vu.lt

H. Gholami et al. (eds.), *Sustainable Manufacturing in Industry 4.0*,
https://doi.org/10.1007/978-981-19-7218-8_10

orientation comprises software customization services based on each organization's requirements [10].

On the other hand, academic research on Industry 4.0 focuses predominantly on large businesses and just moderately on Small and Medium Enterprises (SMEs). However, many large enterprises collaborate with SMEs [11, 12]. As a result, large enterprises' actions influence the decisions of their smaller partners and their demands and influence the configuration of SMEs concerning the technical advancements resulting from Industry 4.0. Consequently, it is crucial to evaluate how SMEs adopt Industry 4.0 and how the latter affects SMEs' generation of industrial value [13]. On top of that, the future of SMEs, which are vital contributors to most sectors and nations, mainly relies on their ability to adapt to customer expectations while preserving a competitive edge in their market [14]. To this end, SMEs must continuously enhance their industrial management procedures, including planning, resource use, production control, and operational performance measurement and evaluation [15]. For instance, the adoption of Enterprise resource planning (ERP) systems was handled differently in small and medium-sized enterprises (SMEs) than in large enterprises since SMEs frequently have lower levels of digitization than their global businesses, which is likely to impact the implementation of CPS. In addition, other SMEs are involved in specialized markets, delivering small-batch or custom-made items. Frequently, these enterprises require simplified versions of the equipment and instruments utilized in the manufacturing facilities of large enterprises [16]. Therefore, investigating the perspective of SMEs on the feasibility of adopting Industry 4.0 contributes to a complete understanding of the organizational consequences throughout industrial value chains [17].

Besides, Industry 4.0 adoption has gained greater visibility and importance due to its implications for achieving sustainability [18]. Traditional manufacturing methods are infamous for their negative environmental impacts. Traditional industrial processes and technologies are responsible for increased resource use, global warming, severe environmental damage, and pollution [19]. Industry 4.0 may significantly contribute to sustainability by reducing carbon footprints, using renewable energy, and developing technological solutions beneficial for individuals and society [17]. The growth of Industry 4.0 facilitates more transparent resource optimization. By implementing Industry 4.0 techniques, it is possible to increase production efficiency and innovation, affecting environmental and social sustainability [20]. All things considered, Industry 4.0 adoption could benefit SMEs while it is a path to sustainability, motivating the present study to figure out Industry 4.0 adoption indicators for sustainability and propose a Multi-Criteria Decision Making (MCDM) framework to evaluate Lithuanian SMEs concerning the identified indicatros. To this end, a regrious review was conducted, at first, to figure out the indicators. Afterward, a novel integrated CRiteria Importance Through Inter-criteria Correlation (CRITIC)-Technique for Order of Preference by Similarity to Ideal Solution (TOPSIS) method under Fermatean fuzzy sets is proposed to evaluate the Lithuanian SMEs in terms of adopting industry 4.0 for sustainability. The main contributions of the present study are presented below:

1. Finding the indicators of industry 4.0 adoption for achieving sustainability in SMEs and proposing a comprehensive framework of indicators for evaluating the SME's performance.
2. Proposing an assessment framework to evaluate the performance of SMEs based on the identified indicators. Fermatean fuzzy sets have endorsed the present framework to deal with uncertainty and indeterminacy in decision-making.
3. Applying the proposed assessment framework to evaluate five Lithuainain SMEs to study the applicability and efficiency of the proposed framework in dealing with real-life problems.

The present study is organized as follows: The identified indicators were introduced and presented in Sect. 2. The proposed method and the preliminaries of Fermatean fuzzy sets are presented in Sect. 3. The results of applying the proposed method for evaluating the Lithuanian SMEs are presented in Sect. 4 using tables and figures. Section 5 provided a sensitivity analysis to study the sensitivity of the proposed method concerning the weight changes. The results of the present research are discussed in Sect. 6. Moreover, broad conclusions are provided in Sect. 7.

2 Industry 4.0 Adoption Indicators for Sustainability

2.1 Profitability

Industry 4.0's effects on profitability have been extensively recognized. Numerous economic sustainability opportunities have been linked to implementing Industry 4.0 technology advancements, including additive manufacturing, the industrial internet of things (IIoT), data analytics, and cloud service [3]. Industry 4.0 enables organizations to have flexible production processes and analyze vast volumes of data in real-time, enhancing planning and accurate decisions. On top of that, I4.0 could benefit enterprises through resource efficiency, boosting innovation, improving reliability and capacity, and minimizing inventory costs [21]. Also, Industry 4.0 is anticipated to play a crucial role in the transition of industrial and social organizations toward sustainable growth. Industry 4.0 promotes high-efficiency improvements, resulting in enhanced organizational performance across all three aspects of sustainability [22] [23].

2.2 Emissions Reduction

As industrial emissions account for even more than Forty percent of global greenhouse gas emissions, it is considered that the digitalization of production and the rise of industry 4.0 provide various chances to cut carbon emissions [24]. The IIoT and AI-based manufacturing, for instance, boost the efficiency and adaptability of production, decrease waste, and lower the greenhouse gas index per unit. Furthermore, Industries will tend to have been energy-intensive with high carbon dioxide emissions soon. On the other hand, it is anticipated that advanced manufacturing will revolutionize the sector and its associated supply chains and significantly decrease the fossil-carbon footprint per unit of product or product service, thereby assisting societies in achieving environmental and social sustainability [25, 26]. The opportunities provided by Smart manufacturing for the creation of new markets, including the transition from large-scale production to product customization and even device personalization, could improve the marketplace and play a role in the embodiment of a low-carbon future, thereby furthering the ecological sustainability.

2.3 Economic Development

When technological trends and design concepts of Industry 4.0 are embraced across the business ecosystem, it is thought that digitalization may significantly contribute to the sustained economic development of nations. Industry 4.0 is anticipated to create jobs rather than eliminate them [27]. Even though Industry 4.0 eliminates several low-skilled occupations, it generates limitless digitalization-related employment prospects. Industry 4.0 might enable less-developed nations to leapfrog their unrealized industrial progress and hasten the modernization of their economies [28, 29]. Industry 4.0 is anticipated to have additional effects on management and future employment, enabling the creation of new business models that will have a significant impact on industry and sectors of the economy, ultimately affecting the entire product lifecycle, developing a different way of producing goods and conducting business, allowing the optimization of processes and boosting the company's competitiveness [4, 30]. In contrast, the growth of the circular economy is aided by the sustainability of operations management brought about by the use of Industry 4.0 digitalization [31]. Given that the ramifications of Industry 4.0 extend beyond the supply chain or industry borders and encompass distribution channels and the marketplace, the proliferation of smart manufacturing will present good chances for different facets of sustainable economic growth [32].

2.4 Business Model Innovation

Industry 4.0's implications for business elements allow identifying several transformation strategies for obsolete models. First, an enhancement of the conventional business model through incremental innovation in value generation, as well as value delivery, has already been outlined; Second, radical innovation has been defined as diversifying the actual business model by reconfiguring value-networked ecosystems; third, a new business model paradigm founded on the amortization of goods and services has been presented [33, 34]. Furthermore, industry 4.0 has a significant impact on business models due to the fact that this new production paradigm requires a new mode of communication throughout supply chains. Industry 4.0 entails the existence of a comprehensive communication network linking businesses, suppliers, resources, logistics, factories, and customers [35]. Each sector improves its configuration in real-time based on the needs and conditions of related sections in the network, maximizing the revenue for all cooperatives with little resource sharing. The issue of business models is resolved in light of how a company's positioning aids in comprehending how to earn profits from Industry 4.0. Positioning as an Industry 4.0 user and provider influences SME business models [4].

2.5 Human Resource Development

Industry 4.0 and digital transformation are significantly altering human resources' working ways. Experts think that system simplification, automation, and improved decision-making may considerably increase human resource productivity [36]. AI and data analytics technologies, for instance, can help managers identify relevant patterns from employee data and offer customized career development or educational methods based on every employee's behavior, experience, talents, personality, and learning habits. Using IoP in a business environment, also known as social intranets, enables employers and employees to connect more openly and interactively, closing the communication gap between leaders, managerial levels, and employees [29]. Moreover, graphical and simulation systems such as AVR provide one of the most efficient methods for industrial training. AVR provides a cheaper, safer, faster, and more effective learning environment. Before committing to complex or delicate repairs, maintenance professionals, for instance, might rehearse them safely and increase their preparedness [37]. In addition, organizations may use AI and advanced analytics to examine the background of a particular employment post and select the most qualified individuals with the necessary skills from the pool of current talent. In turn, digitally connected human resource development programs provide several chances for socioeconomic sustainability, such as staff productivity and overall business efficiency [38, 39].

2.6 Sustainable Resources Development

Industry 4.0's digital revolution promotes environmental sustainability through resource transformation and sustainable energy. Industry 4.0 radically revolutionizes civilizations' production, commerce, consumption, and lifestyle [40]. Digitalizing energy technologies, such as wireless networks and blockchain technology, has created significant prospects for the sector's advancement [41]. Smart grids facilitating the convergence of power networks and renewable technologies are an example of generally accepted digitalization implications [29]. concluded that Industry 4.0's contribution to energy sustainability begins with digitizing the supply side of the energy sector, specifically by reducing the operational and maintenance costs of energy plants and energy generation facilities, enhancing the efficiency and safety of energy delivery networks, and enhancing the overall visibility and control of energy production and delivery operations. Sustainable energy development is only one of the benefits of Industry 4.0 adoption. Also, efficient allocation and production systems, advanced digital manufacturing techniques, and smart material planning have considerably increased material efficiency and cost savings, opening the road for sustainable development [42]. Furthermore, decentralization, interoperability, consolidation, and real-time capacity of Industry 4.0 have significant time efficiency implications for manufacturing cycle time [21].

2.7 *Efficiency and Productivity*

Industry 4.0's effects on manufacturing productivity and efficiency are well-documented. Industry 4.0's digitalization of production permits developing and implementing a hybrid systems engineering ecosystem to endorse the product personalization philosophy [43]. Consequently, the asynchronous manufacturing capabilities of the lean supply chain would enable mass-production-capable production facilities to successfully meet ever-changing client demands, even though they are of small-batch or even a single-item production [44, 45]. Alternately, automation and interoperability assist in manufacturing efficiency and productivity by allowing essential maintenance, monitoring machine effectiveness, enhancing scheduling efficiency, and decreasing machine downtime. Moreover, industrial automation decreases human interference, resulting in fewer human mistakes, decreased risk, and fewer safety problems [46]. On top of that, it is concluded that industry 4.0 enhances flexibility and resource efficiency by turning disconnected and manual manufacturing operations into interconnected and digitalized, interoperable systems in an innovative environment that can enable decision-making across massive actual data, tangible interactions, and cooperation with machines, sensors, and operators, thereby enhancing decision-making processes and accelerating collaboration [29, 47].

2.8 *Ecological Responsibility*

Owing to the emergence of proactive and reactive environmentally friendly practices, Industrial 4.0 and the digitalization of the industrial sector have significant consequences for socioeconomic sustainability [48]. For instance, additive manufacturing technologies enable the creation of new eco-friendly items. Eco Balance, Life cycle assessment, and eco-efficiency benchmarking are information-intensive environmental management approaches [49]. Industry 4.0, directed by smart technologies and manufacturing mechanisms, would potentially minimize industrial waste, overproduction, goods mobility, and energy use [26]. From the economic sustainability perspective, the digital revolution would enable enterprises to acquire market knowledge and other environmental sustainability prospects [22]. Regarding waste reduction and material efficiency, the productivity effects of Industry 4.0, empowered by production flexibility, collaborative production management, design modularity, and supply chain-wide knowledge management capabilities, present numerous opportunities for environmental sustainability [50].

2.9 *Flexible and Agile Production*

Today's manufacturers face demand instabilities, product customization requirements, and decreased product and production technology longevity. Industry 4.0 aids enterprises' sustainability under these conditions by allowing enterprises to establish a more flexible and agile production system [51]. Intelligent Next-Generation Enterprise Resource Planning (ERP), big data analytics, digital twins, and industrial simulation allow businesses to effectively manage environmental uncertainties, micromanage change processes, and transform their current business model(s) in a turbulent business environment [52]. In the context of Industry 4.0, the digitalized virtual intimacy and cloud-based ability to connect across value chains, intelligent production plants, and decentralization would establish an agile and lean manufacturing ecosystem that allows rapid reaction and adaptation strengths in response to adapt and environmental uncertainties [53, 54].

2.10 *Social Welfare Improvement*

Numerous employment options and an increase in minimum earnings due to the skill-intensive nature of new professions in the context of Industry 4.0 can effectively combat economic inequality [55]. In addition, the new marketing and distribution strategies and the resources, materials, and production efficiency provided by smart-digitized manufacturing are anticipated to increase the worldwide availability and affordability of goods and services [56]. Furthermore, Industry 4.0 and the new paradigm of digitalized production can provide excellent chances to reduce income and wealth inequality [57]. Numerous employment options and an increase in minimum earnings due to the skill-intensive nature of new professions in Industry 4.0 can effectively combat economic inequality [58]. Moreover, Business model innovation within Industry 4.0, the PaaS model, in particular is altering the notion of ownership by diminishing the value of prized assets and facilitating the affordability of commodities when required [22]. In other words, digital transformation is integrating and utilizing digital technology to boost production and social welfare. Beyond products, digitalization impacts the business model, management systems, organization, and the entire value chain [59].

2.11 Job Creation

According to studies, industry 4.0 substantially influences the recruiting industry. In Industry 4.0, robotic systems, autonomous machines, and intelligent devices replace people in various tasks, including inventory monitoring, quality management, and product distribution [28, 60]. Furthermore, Industry 4.0 technologies support the TBL ideas through enhanced productivity, monitoring of energy usage, reduced resource consumption, a secure work environment, higher staff satisfaction, and new employment development [61]. Moreover, collaboration in public policy is required to investigate the future environmental and social benefits of digital technology. In other words, Industry 4.0 and the digitalization of the manufacturing sector lead to a sustainable future economy, creating many sustainable manufacturing-related jobs [29]. Besides, scholars anticipate that Industry 4.0 will destroy a sizable part of low- to middle-skilled positions and counteract the automation-related job loss by creating countless new job openings in informatics, system integration, process engineering, and mechatronics [62]. The social and environmental sustainability consequences of Industry 4.0 are also not confined to generating job possibilities associated with digitalization [29].

2.12 Manufacturing Cost Reduction

By incorporating existing technologies to support a more viable approach for Industry 4.0 requirements, technology companies could indeed reduce these costs while gaining access to a more significant segment of consumers, thereby boosting their revenues and reaching marginal cost reduction, which requires a more extensive convergence of technological developments and standardized solutions [21, 63, 64]. Industry 4.0's cost-saving benefits for the manufacturing sector are extensively studied. The improved process controllability, accident prevention and real-time monitoring, non-stop production, improved manufacturing quality and precision, maintenance efficiency, lower human errors, higher equipment effectiveness, streamlined procurement processes, quality decision-making, material/resource/energy efficiency, and reduced human resource costs are some Industry 4.0 advantages for reducing the manufacturing cost [65–67].

2.13 Modularized Production

Numerous businesses are attracted to Industry 4.0 due to the promise of personalized and customized manufacturing at the exact cost of mass production [68]. Industry 4.0 and the underpinning intelligent digital technologies can promote sustainability by empowering firms to adopt a modular system for manufacturing processes, engineering, production, and distribution [69]. Moreover, based on I4.0, intelligent machines and elements in a digital network generate a so-called cyber–physical platform that supports modular and variable manufacturing as a requirement for unique single-item production [70, 71]. Also, the modular product design provides benefits such as a quicker time to market, decreased production costs and complexity, increased product quality, a longer operational lifetime, and energy and material economy [72]. Furthermore, the supportive involvement of Cyber–Physical Production Systems (CPPS) and IIoT and interoperability would enable the physical modularization of manufacturing equipment, infrastructures, or entire production networks, conditions that allow production facilities to be easily converted and utilized for substitute processes and technologies without high reconfiguration and automation costs [73]. Production modularization offers further sustainability benefits, including increased productivity, enhanced process stability, product customization, and decreased waste and lead times [74].

2.14 Security Improvement

The consequences of Industry 4.0 on risk management and planning are varied. End-to-End (E2E) visibility will increase as a result of the implementation of IIoT, semantic technology, cloud data, and advanced analytics, as well as the elimination of information silos and the streamlined flow of data regarding plant capacity, inventory level, procurement schedules, machine conditions, and transportation routes [26, 75]. In turn, the data-driven E2E accessibility leads to a decrease in manufacturing risk and improved stability. In addition, many Industry 4.0-related advancements incorporate sophisticated safety precautions, such as open SAFETY, for manufacturing machines' safe and dependable running. Industry 4.0-compatible solutions for maintenance management that enable autonomous and real-time troubleshooting and problem-solving of assets greatly minimize safety concerns in dynamic production situations [46, 76]. Consequently, Industry 4.0 enables producers to recognize possible problems in real-time and take preventative measures before they become actual threats. Smart safety wearables Intelligent cameras, AI-based location awareness systems, and smart sensors may identify and report any human or machine activity that poses a safety danger. Industry 4.0 has also been linked to the rising use of safer and more smart Collaborative Robots (cobots) in intelligent factories [29]. As a result of improvements in AI, machine learning, data analytics, and smarter cobots now have enhanced danger recognition and risk evaluation capabilities. Intelligent cobots better perceive the world surrounding them, decrease operational risk, and make the human workforce safer [77, 78].

2.15 Personalization

Industry 4.0 offers new possibilities for product customization. Manufacturers may remain competitive in the era of digitalization by using the production customization strategy, the most recent differentiation strategy [79]. Industry 4.0 improves manufacturing processes' adaptability to construct items with a high degree of personalization, similar to the artisanal manufacturing age. This circumstance is particularly pertinent to SMBs owing to their agility. Industry 4.0 helps manufacturers adapt to increased customization since buyers value individualized services and goods more than conventional offers. By focusing on high-margin items and production process innovation, SMEs may maintain their competitive edge over multinational rivals by producing customized products with short lead times [80]. The current availability of additive manufacturing and smart factory production flexibility enables manufacturers to create Ultra-Personalized Products (UPPs) depending on customer preferences and novel concepts. As a result of introducing flexible large-scale production systems, customers may obtain UPPs at a much more reasonable cost, and producers can earn a more significant profit per product unit [80]. Moreover, data mining technologies have represented a growing to engage and interact with customers directly and gather and analyze a vast quantity of data about client preferences and consumption patterns [81]. Also, AI-based production planning and digital twin models contribute to the optimal synchronization of concurrently created UPPs' manufacturing practices and operations at the production level [82].

2.16 Supply Chain Digitization and Integration

Traditional supply chains are transformed into the Digital Supply Network (DSN) due to technological advancements such as IIoT, Blockchain, cloud computing, and sophisticated analytics [29, 83]. Ghobakhloo [29] mentioned that DSN consists of three distinct functional levels. At the physical-digital layer of the value network, signals are gathered using smart technology, machine vision, and actuators. Machine and process controllers convert physical world signals into meaningful digital data, including the control system and control systems. Artificial intelligence (AI) and business analytics capabilities in most current ERP packages provide actionable insights from digital records at the digital-digital layer. Tangible resources across the supply network autonomously execute AI-driven choices based on the digital record at the digital-physical layer. The real-time, dynamic, integrative, intelligence, scalable, and agility characteristics of DSNs provide numerous benefits, including deeper customer integration, financial flow integration, increased operational efficiency, marketing effectiveness, ad hoc dynamic planning, collaborative planning, collaborative product design, and supply chain-wide workload equality [66, 75]. In addition, DSNs may dramatically reduce digital waste and provide supply members with competitive differentiation owing to improved data gathering and administration, information integration, and physical process execution [67].

3 Methodology

The present study evaluates five Lithuanian SMEs' performance in adopting industry 4.0 for sustainability. To this end, The CRITIC method, proposed by Diakoulaki et al. [84], is applied to compute the objective weight of indicators. Afterward, the TOPSIS method evaluates the SMEs concerning the identified indicators. Several studies integrated these two methods to deal with MCDM problems under various types of fuzzy environments. For instance [85], integrated these two methods to select a hybrid REs System for households using crisp numbers. Xu et al. [86] combined these two methods to select a set of Pareto solutions determined by a novel non-dominated sorting genetic algorithm [87]. integrated these two methods to assess low-cost airlines' financial performance and service quality [88]. combined these two methods to evaluate the financial performance of Serbian banks using crisp numbers. Also, Adalı and Tuş [89] integrated CRITIC with several MCDM methods, including TOPSIS, to rank hospital sites using crisp numbers.

However, the present study integrated CRITIC and TOPSIS under Fermatean fuzzy sets to deal with uncertainty and indeterminacy in decision making. Fuzzy sets (FSs) theory has been widely utilized by scholars to deal with uncertainty and imprecision. Atanassov [90] developed the concept of Intuitionistic fuzzy sets (IFSs), an extension of the standard fuzzy theory in which the sum of belongingness degree (BD) and non-belongingness degree (NBD) should always be less than or equal to one. After that, Yager [91] suggested Pythagorean fuzzy sets (PFSs) to address the constraints of IFSs and deal with ambiguity, uncertainty, and imprecision in real-world applications. Numerous researchers have applied PFSs to various fields, including diagnosing diseases, managing hospital waste, sustainable supplier assessment, evaluating pharmacological therapy, and improving pattern recognition [92]. Nevertheless, whenever the squared summation of BD. and NBD is more than one, Pythagorean fuzzy sets are inapplicable, which prompted [93] to design Fermatean fuzzy sets (FFSs) to overcome the problem mentioned above. In particular, the cube total of BD. and NBD is below or equal to one in FFSs, allowing FFSs more capable of adequately addressing complicated MCDA concerns.

Regarding the experts, the present study asked three experts' opinions concerning the five Lithuanian SMEs based on identified indicators. To this end, linguistic variables are used shown in Table 1.

Table 1 FFNs. and equivalent linguistic terms

Linguistic terms	FFNs
Medium–Low (ML)	(0.4, 0.5, 0.93)
Low (L)	(0.25, 0.6, 0.92)
Very Low (VL)	(0.1, 0.75, 0.83)
Extremely Low (EL)	(0.1, 0.9, 0.65)
Medium (M)	(0.5, 0.4, 0.93)
Extremely High (EH)	(0.9, 0.1, 0.65)
Very High (VH)	(0.8, 0.1, 0.79)
High (H)	(0.7, 0.2, 0.87)
Medium–High (MH)	(0.6, 0.3, 0.91)

3.1 Preliminaries

Definition 1. Senapati and Yager [93] Let Δ a restricted universe of discourse; consequently, the first equation is shown as a fermatean fuzzy set.

$$F = \{\langle f_i, (b_F(f_i), n_F(f_i))\rangle | f_i \in \Delta\} \tag{1}$$

In which $b_F : \Delta \to [0.1]$ indicates the belonging degree of the element $f_i \in \Delta$ in an FFS, and $b_F : \Delta \to [0.1]$ shows the non-belonging degree of the element $f_i \in \Delta$ in an FFS. Also, the condition $0 \leq \left(b_f(f_i)\right)^3 + (n_F(f_i))^3 \leq 1$ must be fulfilled for each $f_i \in \Delta$.

Definition 2. Senapati and Yager [93] Assume $\gamma = \left(b_\gamma, n_\gamma\right)$, the indeterminacy degree of an FFS is shown by Eq. 2.

$$\pi_\gamma = \sqrt[3]{1 - b_\gamma^3 - n_\gamma^3} \tag{2}$$

where b_γ *and* $n_\gamma \in [0, 1]$, and $0 \leq b_\gamma^3 + n_\gamma^3 \leq 1$.

Definition 3. Senapati and Yager [93] Let $\lambda = (b_\lambda, n_\lambda)$ an FFS, the score and accuracy functions of γ is computing by Eqs. 3 and 4

$$F_s(\gamma) = b_\gamma^3 - n_\gamma^3 | -1 \leq F_s(\gamma) \leq 1 \tag{3}$$

$$F_a(\gamma) = b_\lambda^3 + n_\gamma^3 | 0 \leq F_a(\lambda\gamma) \leq 1 \tag{4}$$

Moreover, The following comparable methods may be employed to rank $\gamma_1 = \left(b_{\gamma_1}, n_{\gamma_1}\right)$ and $\gamma_2 = \left(b_{\gamma_2}, n_{\gamma_2}\right)$.

(a) If $F_s(\gamma_1) > F_s(\gamma_2)$ then $\gamma_1 > \gamma_2$
(b) If $F_s(\gamma_1) < F_s(\gamma_2)$ then $\gamma_1 < \gamma_2$

(c) If $F(\gamma_1) = F_s(\gamma_2)$ then,

I. If $F_a(\gamma_1) > F_a(\gamma_2)$ then $\gamma_1 > \gamma_2$.
II. If $F_a(\gamma_1) < F_a(\gamma_2)$ then $\gamma_1 < \gamma$.
III. If $F_a(\gamma_1) = F_a(\gamma_2)$ then $\gamma_1 = \gamma$.

Definition 4. Senapati and Yager [93] assume $\gamma = (b_\gamma, n_\gamma), \gamma_1 = (b_{\gamma_1}, n_{\gamma_1})$ and $\gamma_2 = (b_{\gamma_2}, n_{\gamma_2})$ are three FFSs. Some operators for FFSs are presented using equations five to eleven.

$$\gamma^c = (n_\gamma, b_\gamma) \tag{5}$$

$$\gamma_1 \cap \gamma_2 = (min\{b_{\gamma_1}, b_{\gamma_2}\}, max\{n_{\gamma_1}, n_{\gamma_2}\}) \tag{6}$$

$$\gamma_1 \cup \gamma_2 = (max\{b_{\gamma_1}, b_{\gamma_2}\}, min\{n_{\gamma_1}, n_{\gamma_2}\}) \tag{7}$$

$$\gamma_1 \oplus \gamma_2 = \left(\sqrt[3]{b_{\gamma_1}^3 + b_{\gamma_2}^3 - b_{\gamma_1}^3 b_{\gamma_2}^3}, n_{\gamma_1} n_{\gamma_2}\right) \tag{8}$$

$$\gamma_1 \otimes \gamma_2 = \left(b_{\gamma_1} b_{\gamma_2}, \sqrt[3]{n_{\gamma_1}^3 + n_{\gamma_2}^3 - n_{\gamma_1}^3 n_{\gamma_2}^3}\right) \tag{9}$$

$$l\gamma = \left(\sqrt[3]{1 - (1 - b_\gamma^3)^l}, (n_\gamma)^l\right), l > 0 \tag{10}$$

$$\gamma^l = \left((b_\gamma)^l, \sqrt[3]{1 - (1 - n_\gamma^3)^l}\right), l > 0 \tag{11}$$

3.2 *Proposed FF-CRITIC-TOPSIS*

Step 1. A matrix for experts' opinions

Assume that, $\{E_1, E_2, \ldots, E_m\}$ be a set of SMEs, $\{I_1, I_2, \ldots, I_n\}$ a set of indicators, and $\{DE_1, DE_2, \ldots, DE_p\}$ a group of Decision Experts, supporting each enterprise E_i concerning an indicator I_j using Fermatean fuzzy linguistic variables. The decision matrix (D) is $D = \left(o_{ij}^k\right), fori = i, \ldots, m; j = 1, \ldots, n$, while o_{ij}^k indicates the given idea regarding the enterprise (i) concerning the indicator (j) by kth decision experts [94].

Step 2. Experts' significance

FFNs present DEs' significances; then $\omega_k = (b_k, n_k)$ is the importance of kth decision expert. Equation 12 is used to calculate experts' significance.

$$\omega_k = \frac{\left(b_k^3 + \pi_k^3 \times \left(\frac{b_k^3}{b_k^3 + n_k^3}\right)\right)}{\sum_{k=1}^{p}\left(b_k^3 + \pi_k^3 \times \left(\frac{b_k^3}{b_k^3 + n_k^3}\right)\right)}, K = 1, \ldots, p; \omega_k \geq 0, \sum_{1}^{p} \omega_k = 1. \quad (12)$$

Step 3. Aggragation

Fermatean fuzzy weighted averaging (FFWA) operator aggregates individual decision matrices. Let $\mathring{A} = (a_{ij})_{m\times n}$ be the aggregated FF-decision matrix, where.

$$a_{ij} = \left(\sqrt[3]{1 - \prod_{k=1}^{r}\left(1 - \left(b_{ij}^k\right)^3\right)^{\omega_k}}, \prod_{k=1}^{r}\left(n_{ij}^k\right)^{\omega_k}\right) \quad (13)$$

Step 4. CRITIC

Step 4.1. A matrix for scores

Constructing a matrix, $\Lambda = (k_{ij})_{m\times n}$, for scores is the first step of CRITIC, constructed by Eq. 14.

$$k_{ij} = \frac{1}{2}\left[\left(b_j^3 - n_j^3 - \ln\left(1 + \pi_j^3\right)\right) + 1\right], \text{ for } i = 1, \ \ldots, \ m \quad (14)$$

Step 4.2. Normalization

The normalized score matrix $\tilde{\Lambda} = \left(\tilde{k}_{ij}\right)_{m\times n}$ is constructed by Eq. 15.

$$\tilde{k}_{ij} = \begin{cases} \frac{k_{ij} - k_j^-}{k_j^+ - k_j^-}, & j \in N_b \\ \frac{k_j^+ - k_{ij}}{k_j^+ - k_j^-}, & j \in N_n \end{cases} \quad (15)$$

where $k_j^- = \min_i k_{ij}$ and $k_j^+ = \max_i k_{ij}$.

Step 4.3. Standard deviation

Standard deviations are computed by Eq. 16.

$$\sigma_j = \sqrt{\frac{\sum_{i=1}^{m}\left(\tilde{k}_{ij} - \overline{k}_j\right)^2}{m}} \quad (16)$$

Step 4.4. Correlation

Correlations are calculated by Eq. 17.

$$r_{jt} = \frac{\sum_{i=1}^{m}\left(\tilde{k}_{ij} - \overline{\mathrm{k}}_j\right)\left(\tilde{k}_{it} - \overline{\mathrm{k}}_t\right)}{\sqrt{\sum_{i=1}^{m}\left(\tilde{k}_{ij} - \overline{\mathrm{k}}_j\right)^2 \sum_{i=1}^{m}\left(\tilde{k}_{it} - \overline{\mathrm{k}}_t\right)^2}} \tag{17}$$

Step 4.5. Information quantity

The information quantity is computed by Eq. 18.

$$\nu_j = \sigma_j\left(\sum_{t=1}^{n}(1 - r_{jt})\right) \tag{18}$$

Step 4.6. Weight determination

Equation 19 is used to determine the weight of indicators.

$$\varpi_j = \frac{\nu_j}{\sum_{i=1}^{m}\nu_j} \tag{19}$$

Step 5. Fermatean fuzzy positive and negative ideal solutions

$$\begin{aligned} S^+ &= \left\{\left(\max_i k_{ij} | j \in J\right), \left(\min_i k_{ij} | j \in j'\right) | \; i = 1, \ldots, m\right\} \\ &= \left\{k_1^+, \, k_2^+, \, \ldots, \, k_n^+\right\} \end{aligned} \tag{20}$$

$$\begin{aligned} S^- &= \left\{\left(\min_i k_{ij} | j \in J\right), \left(\max_i k_{ij} | j \in J'\right) | \; i = 1, \ldots, m\right\} \\ &= \left\{k_1^-, \, k_2^-, \, \ldots, \, k_n^-\right\} \end{aligned} \tag{21}$$

Step 6. Relative closeness

Relative closeness to the fermatean fuzzy ideal solutions is calculated by Eq. 22 [95].

$$R(K_i) = \frac{Y_i^-}{Y_i^- + Y_i^+} \quad \text{for } i = 1, \ldots, m$$

where

$$\begin{aligned} Y_i^- &= dis\left(S^-, z_{ij}\right) = \varpi_j\sqrt{\frac{1}{2}\left[\left((b_{ij})^3 - \left(b_j^-\right)^3\right)^2 + \left((n_{ij})^3 - \left(n_j^-\right)^3\right)^2 + \left((\pi_{ij})^3 - \left(\pi_j^-\right)^3\right)^2\right]} \\ Y_i^+ &= dis\left(S^+, z_{ij}\right) = \varpi_j\sqrt{\frac{1}{2}\left[\left((b_{ij})^3 - \left(b_j^+\right)^3\right)^2 + \left((n_{ij})^3 - \left(n_j^+\right)^3\right)^2 + \left((\pi_{ij})^3 - \left(\pi_j^+\right)^3\right)^2\right]} \end{aligned} \tag{22}$$

4 Results

In the first step, the decision-making matrix should be constructed. It is assumed that the importance of the first expert is $(0.8, 0.45, 0.67)$, for the second one is $(0.75, 0.55, 0.74)$, and the third one is $(0.8, 0.5, 0.71)$. Table 2 shows the decision-making matrix.

After determining the significance of the DEs, the aggregated decision-making matrix must be generated, as depicted in Table 3.

Next, it is necessary to generate and normalize the score matrix. In Table 4, the normalized score matrix is displayed. In addition, Table 4 displays the standard deviation, the amount of data, and the weight of the indicators.

Table 5 indicates the positive and negative ideal solutions.

The relative closeness and result ranking of the enterprises are displayed in Table 6.

According to the results, the preference order of the SEMs concerning the identified industry 4.0 adoption indicators for sustainability is $E_5 > E_2 > E_4 > E_3 > E_1$.

5 Sensitivity Analysis

This section determines the weight of various indicators and assesses the method's sensitivity to the weight change. In other words, n sets of indicator weights must be specified for the sensitivity analysis if there are n indicators. Each set has the most critical (weighted) and least important (unweighted) criteria, whilst others contain weights between the most and least important criteria [96]. The preceding stage should generate sixteen sets of indicator weights for the sensitivity analysis. The proportional value of each indicator in each set is displayed in Table 7.

SMEs were also ranked for each set after determining the weight of indicators in each set. Figure 1 displays the outcomes of the sensitivity analysis. According to Fig. 1, the proposed method is weight dependent, yet E_5 is the optimal enterprise in most cases.

6 Discussion

According to the results, the most critical indicator is "flexible and agile production." Generally speaking, by implementing a new technology, Industry 4.0 may help manufacturing organizations achieve cost-efficiency and flexibility in industrial manufacturing processes. Agile manufacturing establishes a real-time production system that can rapidly switch between product models or business units in

Table 2 Decision-making matrix

		E1	E2	E3	E4	E5
I1	DE1	MH	H	ML	L	EH
	DE2	ML	M	ML	MH	H
	DE3	L	L	ML	ML	M
I2	DE1	MH	ML	L	MH	MH
	DE2	H	MH	ML	VL	H
	DE3	MH	M	L	M	MH
I3	DE1	L	MH	ML	H	MH
	DE2	MH	ML	L	VH	H
	DE3	MH	H	ML	L	H
I4	DE1	L	H	M	H	H
	DE2	M	H	VL	ML	MH
	DE3	L	MH	M	M	M
I5	DE1	H	ML	L	EH	VH
	DE2	ML	ML	M	M	MH
	DE3	EL	MH	MH	ML	L
I6	DE1	M	M	VL	L	M
	DE2	M	ML	ML	L	L
	DE3	L	ML	ML	ML	VH
I7	DE1	VL	MH	H	ML	L
	DE2	ML	ML	M	L	MH
	DE3	ML	H	H	L	M
I8	DE1	M	MH	H	M	H
	DE2	ML	VH	H	MH	H
	DE3	ML	MH	MH	ML	H
I9	DE1	ML	MH	H	L	VL
	DE2	MH	MH	ML	VH	ML
	DE3	MH	ML	MH	L	L
I10	DE1	ML	M	MH	MH	L
	DE2	H	MH	ML	EH	M
	DE3	ML	ML	H	ML	VL
I11	DE1	ML	ML	ML	M	ML
	DE2	H	MH	ML	M	H
	DE3	H	ML	MH	MH	M
I12	DE1	M	MH	ML	H	ML
	DE2	VH	M	M	M	MH
	DE3	ML	ML	M	MH	L

(continued)

Table 2 (continued)

		E1	E2	E3	E4	E5
I13	DE1	ML	ML	H	MH	ML
	DE2	VL	MH	VL	M	ML
	DE3	EL	H	L	ML	ML
I14	DE1	MH	M	ML	M	H
	DE2	VH	MH	ML	M	H
	DE3	EH	MH	M	ML	H
I15	DE1	MH	M	H	EH	ML
	DE2	VL	MH	H	H	L
	DE3	VL	H	ML	MH	ML
I16	DE1	H	MH	H	ML	VH
	DE2	H	H	VH	L	MH
	DE3	MH	MH	VH	ML	M

Table 3 Aggregated decision-making matrix

	E1	E2	E3	E4	E5
I1	γ(0.47,0.45,0.93)	γ(0.56,0.37,0.92)	γ(0.40,0.50,0.93)	γ(0.46,0.45,0.93)	γ(0.77,0.20,0.81)
I2	γ(0.64,0.26,0.90)	γ(0.51,0.39,0.93)	γ(0.31,0.57,0.92)	γ(0.49,0.44,0.93)	γ(0.64,0.26,0.90)
I3	γ(0.54,0.38,0.93)	γ(0.60,0.31,0.91)	γ(0.37,0.53,0.93)	γ(0.68,0.24,0.88)	γ(0.67,0.23,0.88)
I4	γ(0.37,0.41,0.96)	γ(0.67,0.20,0.88)	γ(0.44,0.53,0.92)	γ(0.57,0.39,0.91)	γ(0.61,0.23,0.91)
I5	γ(0.53,0.45,0.91)	γ(0.49,0.42,0.93)	γ(0.50,0.41,0.93)	γ(0.73,0.27,0.84)	γ(0.65,0.26,0.89)
I6	γ(0.44,0.46,0.93)	γ(0.44,0.46,0.93)	γ(0.35,0.57,0.92)	γ(0.32,0.56,0.92)	γ(0.64,0.28,0.90)
I7	γ(0.35,0.57,0.92)	γ(0.60,0.31,0.91)	γ(0.66,0.25,0.89)	γ(0.32,0.56,0.92)	γ(0.49,0.42,0.93)
I8	γ(0.44,0.50,0.92)	γ(0.69,0.15,0.88)	γ(0.67,0.23,0.88)	γ(0.51,0.36,0.94)	γ(0.70,0.20,0.87)
I9	γ(0.55,0.36,0.92)	γ(0.55,0.36,0.92)	γ(0.60,0.31,0.91)	γ(0.59,0.34,0.91)	γ(0.30,0.61,0.91)
I10	γ(0.55,0.38,0.92)	γ(0.51,0.40,0.93)	γ(0.60,0.31,0.91)	γ(0.74,0.25,0.83)	γ(0.36,0.57,0.92)
I11	γ(0.64,0.27,0.90)	γ(0.49,0.43,0.93)	γ(0.49,0.42,0.93)	γ(0.54,0.36,0.93)	γ(0.57,0.35,0.92)
I12	γ(0.63,0.28,0.90)	γ(0.51,0.39,0.93)	γ(0.47,0.43,0.93)	γ(0.62,0.29,0.90)	γ(0.46,0.45,0.93)
I13	γ(0.28,0.70,0.86)	γ(0.60,0.31,0.91)	γ(0.52,0.44,0.92)	γ(0.51,0.39,0.93)	γ(0.40,0.50,0.93)
I14	γ(0.81,0.14,0.77)	γ(0.57,0.33,0.92)	γ(0.44,0.46,0.93)	γ(0.47,0.43,0.93)	γ(0.70,0.20,0.87)
I15	γ(0.43,0.55,0.91)	γ(0.62,0.29,0.90)	γ(0.64,0.28,0.90)	γ(0.78,0.18,0.80)	γ(0.37,0.53,0.93)
I16	γ(0.67,0.23,0.88)	γ(0.64,0.23,0.90)	γ(0.77,0.10,0.81)	γ(0.37,0.56,0.92)	γ(0.67,0.33,0.87)

response to the customer's product type and quantity requirements. Agile manufacturing, unlike lean manufacturing, prioritizes flexibility over cheap cost. For agile manufacturing to implement a corporate-wide adaptable strategy, delivery, production machinery, workers, and the organization must be flexible. On top of that, agile manufacturing may be a foundation for adopting or even inventing Industry 4.0 technologies, according to another take on the relationship between agile manufacturing

Table 4 CRITIC results

	E1	E2	E3	E4	E5	σ_j	ν_j	ϖ_j	Rank
I1	0.11	0.30	0.00	0.10	1.00	0.252	0.218	0.055	14
I2	0.00	0.49	1.00	0.57	0.00	0.339	0.257	0.065	4
I3	0.44	0.69	0.00	1.00	0.98	0.501	0.238	0.060	11
I4	0.00	1.00	0.06	0.52	0.74	0.378	0.217	0.055	15
I5	0.08	0.00	0.01	1.00	0.58	0.280	0.257	0.065	5
I6	0.29	0.27	0.03	0.00	1.00	0.264	0.223	0.056	13
I7	0.03	0.78	1.00	0.00	0.40	0.362	0.254	0.064	6
I8	0.00	0.94	0.85	0.24	1.00	0.492	0.237	0.060	12
I9	0.77	0.77	1.00	0.95	0.00	0.563	0.311	0.079	1
I10	0.38	0.30	0.54	1.00	0.00	0.361	0.241	0.061	10
I11	1.00	0.00	0.04	0.32	0.48	0.304	0.277	0.070	3
I12	0.00	0.73	0.94	0.11	1.00	0.453	0.245	0.062	7
I13	0.00	1.00	0.65	0.68	0.36	0.435	0.216	0.054	16
I14	1.00	0.24	0.00	0.05	0.58	0.309	0.279	0.070	2
I15	0.05	0.46	0.51	1.00	0.00	0.331	0.243	0.061	9
I16	0.64	0.55	1.00	0.00	0.64	0.456	0.245	0.062	8

Table 5 Positive and negative ideal solutions

	S^+	S^-
I1	γ(0.77,0.20,0.81)	γ(0.40,0.50,0.93)
I2	γ(0.31,0.57,0.92)	γ(0.64,0.26,0.90)
I3	γ(0.68,0.24,0.88)	γ(0.37,0.53,0.93)
I4	γ(0.67,0.20,0.88)	γ(0.37,0.41,0.96)
I5	γ(0.73,0.27,0.84)	γ(0.49,0.42,0.93)
I6	γ(0.64,0.28,0.90)	γ(0.32,0.56,0.92)
I7	γ(0.66,0.25,0.89)	γ(0.32,0.56,0.92)
I8	γ(0.70,0.20,0.87)	γ(0.44,0.50,0.92)
I9	γ(0.60,0.31,0.91)	γ(0.30,0.61,0.91)
I10	γ(0.74,0.25,0.83)	γ(0.36,0.57,0.92)
I11	γ(0.64,0.27,0.90)	γ(0.49,0.43,0.93)
I12	γ(0.46,0.45,0.93)	γ(0.63,0.28,0.90)
I13	γ(0.60,0.31,0.91)	γ(0.28,0.70,0.86)
I14	γ(0.81,0.14,0.77)	γ(0.44,0.46,0.93)
I15	γ(0.78,0.18,0.80)	γ(0.37,0.53,0.93)
I16	γ(0.77,0.10,0.81)	γ(0.37,0.56,0.92)

Table 6 FF-TOPSIS outputs

Companies	Y_i^+	Y_i^-	$R(K_i)$	Rank
E1	0.17	0.09	0.34	5
E2	0.14	0.13	0.49	2
E3	0.14	0.12	0.47	4
E4	0.13	0.13	0.48	3
E5	0.13	0.13	0.52	1

and Industry 4.0. Manufacturing agility could be achieved without affecting the knowledge and skills of the workforce. Substantial utilization of the organization's technological resources necessitates knowledgeable personnel and inventiveness, which is needed for agile production [97, 97].

Furthermore, the present research employed a new fermatean fuzzy CRITIC-TOPSIS method to evaluate Lithuanian SEMs concerning the identified indicators. The results indicated that the proposed method could effectively resolve multi-criteria problems characterized by uncertainty and indeterminacy. The primary benefits of the FF-CRITIC-TOPSIS are presented below.

- The proposed methodology could be applied to problems under conventional, Intuitionistic, and Pythagorean fuzzy sets in addition to Fermatean fuzzy sets.
- FFNs were used to address the ambiguity and uncertainty inherent in multi-criteria problems. Additionally, the degree of indeterminacy was addressed at every level of the suggested approach, which improved its efficiency; thus, the proposed model could effectively deal with complex multi-criteria problems.
- Objective weights were determined using the CRITIC approach to avoid subjectivity, which might be viewed as a benefit over conventional techniques. Consequently, the FF-CRITIC-TOPSIS may produce more precise results than previous approaches.

7 Conclusions

The present study investigated how SMEs adopt industry 4.0 functions to achieve sustainability. For this purpose, a rigorous review was done to determine indicators of Industry 4.0 adoption in SMEs; consequently, a novel assessment method was proposed to evaluate the SMEs concerning the identified indicators. The results indicated that agile and flexible production is the essential indicators; thus, it could be concluded that Industry 4.0 could benefit SMEs by boosting their flexibility and agility through employing digital technologies. It has become more critical during the Covid-19 pandemic due to the digital transformation, which highlights flexibility and agility at both organizational and individual levels, especially regarding the delivery of products and services. On top of that, the proposed method could determine

Table 7 Weight sets

	Set 1	Sets2	Set3	Sset4	Set5	Set6	Set7	Set8	Set9	Set10	Set11	Set12	Set13	Set14	Set15	Set16
I1	0.05	0.06	0.06	0.07	0.05	0.06	0.07	0.06	0.08	0.06	0.06	0.06	0.06	0.05	0.06	0.07
I2	0.07	0.05	0.06	0.06	0.07	0.05	0.06	0.07	0.06	0.08	0.06	0.06	0.06	0.06	0.05	0.06
I3	0.06	0.07	0.05	0.06	0.06	0.07	0.05	0.06	0.07	0.06	0.08	0.06	0.06	0.06	0.06	0.05
I4	0.05	0.06	0.07	0.05	0.06	0.06	0.07	0.05	0.06	0.07	0.06	0.08	0.06	0.06	0.06	0.06
I5	0.06	0.05	0.06	0.07	0.05	0.06	0.06	0.07	0.05	0.06	0.07	0.06	0.08	0.06	0.06	0.06
I6	0.06	0.06	0.05	0.06	0.07	0.05	0.06	0.06	0.07	0.05	0.06	0.07	0.06	0.08	0.06	0.06
I7	0.06	0.06	0.06	0.05	0.06	0.07	0.05	0.06	0.06	0.07	0.05	0.06	0.07	0.06	0.08	0.06
I8	0.06	0.06	0.06	0.06	0.05	0.06	0.07	0.05	0.06	0.06	0.07	0.05	0.06	0.07	0.06	0.08
I9	0.08	0.06	0.06	0.06	0.06	0.05	0.06	0.07	0.05	0.06	0.06	0.07	0.05	0.06	0.07	0.06
I10	0.06	0.08	0.06	0.06	0.06	0.06	0.05	0.06	0.07	0.05	0.06	0.06	0.07	0.05	0.06	0.07
I11	0.07	0.06	0.08	0.06	0.06	0.06	0.06	0.05	0.06	0.07	0.05	0.06	0.06	0.07	0.05	0.06
I12	0.06	0.07	0.06	0.08	0.06	0.06	0.06	0.06	0.05	0.06	0.07	0.05	0.06	0.06	0.07	0.05
I13	0.05	0.06	0.07	0.06	0.08	0.06	0.06	0.06	0.06	0.05	0.06	0.07	0.05	0.06	0.06	0.07
I14	0.07	0.05	0.06	0.07	0.06	0.08	0.06	0.06	0.06	0.06	0.05	0.06	0.07	0.05	0.06	0.06
I15	0.06	0.07	0.05	0.06	0.07	0.06	0.08	0.06	0.06	0.06	0.06	0.05	0.06	0.07	0.05	0.06
I16	0.06	0.06	0.07	0.05	0.06	0.07	0.06	0.08	0.06	0.06	0.06	0.06	0.05	0.06	0.07	0.05

Fig. 1 Sensitivity analysis

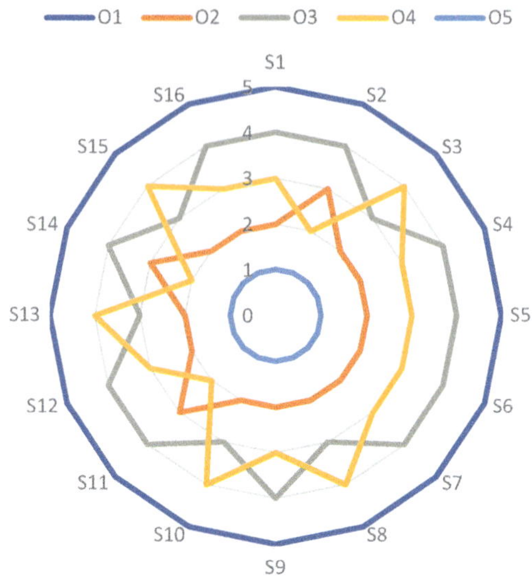

indicators' weight without subjectivity, making the proposed method a practical approach to dealing with various multi-criteria problems in different areas.

The present study has a number of limitations, including, first, the assumption that indicators are independent with no interdependence; second, the experts' unfamiliarity with MCDM as well as how to assist difficulties using linguistic variables made data collection time-consuming. In addition, it is suggested that the proposed methodology be applied in other places for future study and that further expert comments be solicited about the highlighted issues. Also recommended is using the Stepwise Weight Assessment Ratio Analysis (SWARA) approach to determine subjective weights and compare the results to the proposed method.

References

1. Jamwal A, Agrawal R, Sharma M, Giallanza A (2021) Industry 4.0 technologies for manufacturing sustainability: a systematic review and future research directions. Appl Sci 11:5725.
2. Saad SM, Bahadori R, Jafarnejad H (2021) The smart SME technology readiness assessment methodology in the context of industry 4.0. J Manuf Technol Management
3. Hofmann E, Rüsch M (2017) Industry 4.0 and the current status as well as future prospects on logistics. Comput Ind 89:23–34
4. Maresova P, Soukal I, Svobodova L, Hedvicakova M, Javanmardi E, Selamat A, Krejcar O (2018) Consequences of industry 4.0 in business and economics. Economies 6:46.
5. Siltori PF, Anholon R, Rampasso IS, Quelhas OL, Santa-Eulalia LA, Leal Filho W (2021) Industry 4.0 and corporate sustainability: an exploratory analysis of possible impacts in the Brazilian context. Technol Forecast Soc Chang 167:120741.
6. Hermann M, Pentek T, Otto B (2016) Design principles for industrie 4.0 scenarios. In: 2016 49th Hawaii international conference on system sciences (HICSS). IEEE, 3928–3937.

7. Schwab K (2017) The fourth industrial revolution, Currency.
8. Sniderman B, Mahto M, Cotteleer MJ (2016) Industry 4.0 and manufacturing ecosystems: exploring the world of connected enterprises. Deloitte Consult 1:3–14
9. Shahid N, Aneha S (2017) Internet of things: vision, application areas and research challenges. In: 2017 International Conference on I-SMAC (IoT in Social, Mobile, Analytics and Cloud)(I-SMAC). IEEE, 583–587.
10. Buxmann P, Hess T, Ruggaber R (2009) Internet of services. Bus Inf Syst Eng 1:341–342
11. Müller JM, Buliga O, Voigt K-I (2018) Fortune favors the prepared: how SMEs approach business model innovations in industry 4.0. Technol Forecast Soc Chang 132:2–17
12. Stentoft J, Adsboll Wickestrom K, Philipsen K, Haug A (2021) Drivers and barriers for Industry 4.0 readiness and practice: empirical evidence from small and medium-sized manufacturers. Prod Plan & Control 32:811-828
13. Masood T, Sonntag P (2020) Industry 4.0: adoption challenges and benefits for SMEs. Computers in Industry 121:103261.
14. Moeuf A, Pellerin R, Lamouri S, Tamayo-Giraldo S, Barbaray R (2018) The industrial management of SMEs in the era of industry 4.0. Int J Prod Res 56:1118–1136
15. Soni G, Kumar S, Mahto RV, Mangla SK, Mittal M, Lim WM (2022) A decision-making framework for Industry 4.0 technology implementation: the case of FinTech and sustainable supply chain finance for SMEs. Technol Forecast Soc Chang 180:121686.
16. Cimini C, Boffelli A, Lagorio A, Kalchschmidt M, Pinto R (2020) How do industry 4.0 technologies influence organisational change? an empirical analysis of Italian SMEs. J Manuf Technol Manag
17. Kumar R, Singh RK,Dwivedi YK (2020) Application of industry 4.0 technologies in SMEs for ethical and sustainable operations: analysis of challenges. J Clean Prod 275:124063
18. Bai C, Sarkis J (2020) A supply chain transparency and sustainability technology appraisal model for blockchain technology. Int J Prod Res 58:2142–2162
19. Bai C, Dallasega P, Orzes G, Sarkis J (2020) Industry 4.0 technologies assessment: a sustainability perspective. Int J Prod Econ 229: 107776.
20. Bag S, Gupta S, Kumar S (2021) Industry 4.0 adoption and 10R advance manufacturing capabilities for sustainable development. Int J Prod Econ 231: 107844.
21. Dalenogare LS, Benitez GB, Ayala NF, Frank AG (2018) The expected contribution of Industry 4.0 technologies for industrial performance. Int J Prod Econ 204:383–394
22. De Sousa Jabbour ABL, Jabbour CJC, Foropon C, Godinho Filho M (2018) When titans meet–Can industry 4.0 revolutionise the environmentally-sustainable manufacturing wave? The role of critical success factors. Technol Forecast Soc Chang 132:18-25
23. Kamble S, Gunasekaran A, Dhone NC (2020) Industry 4.0 and lean manufacturing practices for sustainable organisational performance in Indian manufacturing companies. Int J Prod Res 58:1319–1337
24. Ford S, Despeisse M (2016) Additive manufacturing and sustainability: an exploratory study of the advantages and challenges. J Clean Prod 137:1573–1587
25. Jin M, Tang R, Ji Y, Liu F, Gao L, Huisingh D (2017) Impact of advanced manufacturing on sustainability: an overview of the special volume on advanced manufacturing for sustainability and low fossil carbon emissions. J Clean Prod 161:69–74
26. Kamble SS, Gunasekaran A, Gawankar SA (2018) Sustainable Industry 4.0 framework: a systematic literature review identifying the current trends and future perspectives. Process Saf Environ Prot 117:408–425
27. Lopes De Sousa Jabbour AB, Jabbour CJC, Godinho Filho M, Roubaud D. (2018) Industry 4.0 and the circular economy: a proposed research agenda and original roadmap for sustainable operations. Ann Oper Res 270:273-286
28. Frey CB, Osborne MA (2017) The future of employment: how susceptible are jobs to computerisation? Technol Forecast Soc Chang 114:254–280
29. Ghobakhloo M (2020) Industry 4.0, digitization, and opportunities for sustainability. J Clean Prod 252:119869.

30. Pereira AC, Romero F (2017) A review of the meanings and the implications of the Industry 4.0 concept. Procedia Manuf 13:1206–1214
31. Cezarino LO, Liboni LB, Stefanelli NO, Oliveria BG, Stocco LC (2019) Diving into emerging economies bottleneck: industry 4.0 and implications for circular economy. Manag Decis.
32. Li L (2018) China's manufacturing locus in 2025: with a comparison of "Made-in-China 2025" and "Industry 4.0." Technol Forecast Soc Chang 135:66–74
33. de Man JC, Strandhagen JO (2017) An Industry 4.0 research agenda for sustainable business models. Procedia Cirp 63:721–726
34. Ibarra D, Ganzarain J, Igartua JI (2018) Business model innovation through industry 4.0: a review. Procedia Manuf 22:4–10
35. Glova J, Sabol T, Vajda V (2014) Business models for the internet of things environment. Procedia Econ Financ 15:1122–1129
36. Margherita EG, Bua I (2021) The role of human resource practices for the development of operator 4.0 in Industry 4.0 organisations: a literature review and a research agenda. Businesses 1:18–33
37. Chinachoti P (2018) The readiness of human resource management for industrial business sector towards industrial 4.0 in Thailand. Asian Adm & Manag Rev 1.
38. Sharma M, Luthra S, Joshi S, Kumar A (2022) Analysing the impact of sustainable human resource management practices and industry 4.0 technologies adoption on employability skills. Int J Manpow
39. Sivathanu B, Pillai R (2018) Smart HR 4.0–how industry 4.0 is disrupting HR, Human Resource Management International Digest.
40. Beier G, Niehoff S, Ziems T, Xue B (2017) Sustainability aspects of a digitalized industry–a comparative study from China and Germany. Int J Precis Eng Manuf-Green Technol 4:227–234
41. Huang Z, Yu H, Peng Z, Feng Y (2017) Planning community energy system in the industry 4.0 era: achievements, challenges and a potential solution. Renew Sustain Energy Rev 78:710–721
42. Kiel D, Muller JM, Arnold C, Voigt K-I (2020) Sustainable industrial value creation: benefits and challenges of industry 4.0. Digit Disruptive Innov. World Scientific.
43. Beier G, Niehoff S, Xue B (2018) More sustainability in industry through industrial internet of things? Appl Sci 8:219
44. Tortorella GL, Fettermann D (2018) Implementation of Industry 4.0 and lean production in Brazilian manufacturing companies. Int J Prod Res 56:2975–2987
45. Venugopal V, Saleeshya P (2019) Manufacturing system sustainability through lean and agile initiatives. Int J Sustain Eng 12:159–173
46. Reis MS, Gins G (2017) Industrial process monitoring in the big data/industry 4.0 era: from detection, to diagnosis, to prognosis. Processes 5:35.
47. Konur S, Lan Y, Thakker D, Morkyani G, Polovina N, Sharp J (2021) Towards design and implementation of Industry 4.0 for food manufacturing. Neural Comput Appl, 1−13.
48. Wang S, Wan J, Zhang D, Li D, Zhang C (2016) Towards smart factory for industry 4.0: a self-organized multi-agent system with big data based feedback and coordination. Comput Netw 101:158–168
49. Wang S, Wan J, Li D, Zhang C (2016b) Implementing smart factory of industrie 4.0: an outlook. Int J Distrib Sens Netw 12:3159805.
50. Niaki MK, Torabi SA, Nonino F (2019) Why manufacturers adopt additive manufacturing technologies: the role of sustainability. J Clean Prod 222:381–392
51. Lasi H, Fettke P, Kemper H-G, Feld T, Hoffmann M (2014) Industry 4.0. Bus Inf Syst Eng 6:239–242
52. Oliveira-Dias D, Maqueira JM, Moyano-Fuentes J (2022) The link between information and digital technologies of industry 4.0 and agile supply chain: mapping current research and establishing new research avenues. Comput & Ind Eng, 108000.
53. Raji IO, Shevtshenko E, Rossi T, Strozzi F (2021) Industry 4.0 technologies as enablers of lean and agile supply chain strategies: an exploratory investigation. Int J Logist Management
54. Singh RK, Kumar P, Chand M (2019) Evaluation of supply chain coordination index in context to Industry 4.0 environment. Benchmarking: Int J

55. Vinuesa R, Azizpour H, Leite I, Balaam M, Dignum V, Domisch S, Felländer A, Langhans SD, Tegmark M, Fuso Nerini F (2020) The role of artificial intelligence in achieving the sustainable development goals. Nat Commun 11:1–10
56. El Baz J, Tiwari S, Akenroye T, Cherrafi A, Derrouiche R (2022) A framework of sustainability drivers and externalities for industry 4.0 technologies using the Best-Worst Method. J Clean Prod, 130909.
57. Choti TM, Kumar S, Yue X, Chan HL (2021) Disruptive technologies and operations management in the Industry 4.0 era and beyond. Prod Oper Manag
58. Xu LD, Xu EL, Li L (2018) Industry 4.0: state of the art and future trends. Int J Prod Res 56:2941–2962
59. Bleicher J, Stanley H (2018) Digitization as a catalyst for business model innovation a three-step approach to facilitating economic success. J Bus Manag 4:62–71
60. Brougham D, Haar J (2018) Smart technology, artificial intelligence, robotics, and algorithms (STARA): employees' perceptions of our future workplace. J Manag Organ 24:239–257
61. Braccini AM, Margherita EG (2018) Exploring organizational sustainability of industry 4.0 under the triple bottom line: the case of a manufacturing company. Sustainability 11:36.
62. Nara EOB, da Costa MB, Baierle IC, Schaefer JL, Benitez GB,Do Santos LMAL, Benitez LB (2021) Expected impact of industry 4.0 technologies on sustainable development: a study in the context of Brazil's plastic industry. Sustain Prod Consum 25:102-122
63. Benitez GB, Ferreira-Lima M, Ayala NF, Frank AG (2021) Industry 4.0 technology provision: the moderating role of supply chain partners to support technology providers. Supply Chain Manag: Int J.
64. Frank AG, Dalenogare LS, Ayala NF (2019) Industry 4.0 technologies: implementation patterns in manufacturing companies. Int J Prod Econ 210:15–26
65. Bechtsis D, Tsolakis N, Vlachos D, Srai JS (2018) Intelligent autonomous vehicles in digital supply chains: a framework for integrating innovations towards sustainable value networks. J Clean Prod 181:60–71
66. Dolgui A, Ivanov D, Sethi SP, Sokolov B (2019) Scheduling in production, supply chain and Industry 4.0 systems by optimal control: fundamentals, state-of-the-art and applications. Int J Prod Res 57:411–432
67. Ivanov D, Dolgui A, Sokolov B (2019) The impact of digital technology and industry 4.0 on the ripple effect and supply chain risk analytics. Int J Prod Res 57:829–846
68. Wang B, Zhao J-Y, Wan Z-G, MA J-H, LI H, MA J (2016a) Lean intelligent production system and value stream practice. Trans Econom Manag (ICEM 2016).
69. Kolberg D, Knobloch J, Zühlke D (2017) Towards a lean automation interface for workstations. Int J Prod Res 55:2845–2856
70. Åkerman M, Fast-Berglund Å, Halvordsson E, Stahre J (2018) Modularized assembly system: a digital innovation hub for the swedish smart industry. Manufacturing Letters 15:143–146
71. Kolberg D, Zühlke D (2015) Lean automation enabled by industry 4.0 technologies. IFAC-PapersOnLine 48:1870–1875
72. Sisinni E, Saifullah A, Han S, Jennehag U, Gidlund M (2018) Industrial internet of things: challenges, opportunities, and directions. IEEE Trans Industr Inf 14:4724–4734
73. Shoval S, Efatmaneshnik M (2019) Managing complexity of assembly with modularity: a cost and benefit analysis. Int J Adv Manuf Technol 105:3815–3828
74. Gupta P (2019) Modularity enablers: a tool for Industry 4.0. Life Cycle Reliab Saf Eng 8:157–163
75. Ardito L, Petruzzelli AM, Panniello U, Garavelli AC (2018) Towards industry 4.0: mapping digital technologies for supply chain management-marketing integration. Bus Process Manag J
76. Braganca S, Costa E, Castellucci I, Arezes PM (2019) A brief overview of the use of collaborative robots in industry 4.0: human role and safety. Occup Environ Saf Health, 641–650.
77. Maurice P, Padois V, Measson Y, Bidaud P (2017) Human-oriented design of collaborative robots. Int J Ind Ergon 57:88–102

78. Schou C, Andersen RS, Chrysostomou D, Bøgh S, Madsen O (2018) Skill-based instruction of collaborative robots in industrial settings. Robot Comput-Integr Manuf 53:72–80
79. Gu X, Koren Y (2018) Manufacturing system architecture for cost-effective mass-individualization. Manufacturing letters 16:44–48
80. Torn I, Vaneker TH (2019) Mass personalization with Industry 4.0 by SMEs: a concept for collaborative networks. Procedia Manuf 28:135–141
81. Yang C, Lan S, Shen W, Huang GQ, Wang X, Lin T (2017) Towards product customization and personalization in IoT-enabled cloud manufacturing. Clust Comput 20:1717–1730
82. Leng J, Liu Q, Ye S, Jing J, Wang Y, Zhang C, Zhang D, Chen X (2020) Digital twin-driven rapid reconfiguration of the automated manufacturing system via an open architecture model. Robot Comput-Integr Manuf 63:101895
83. Wamba SF, Queiroz MM (2022) Industry 4.0 and the supply chain digitalisation: a blockchain diffusion perspective. Prod Plan & Control 33:193–210
84. Diakoulaki D, Mavrotas G, Papayannakis L (1995) Determining objective weights in multiple criteria problems: the critic method. Comput Oper Res 22:763–770
85. Babatunde OM, Munda JL, Hamam Y (2019) Selection of a hybrid renewable energy systems for a low-income household. Sustainability 11:4282
86. Xu C, Ke Y, Li Y, Chu H, Wu Y (2020) Data-driven configuration optimization of an off-grid wind/PV/hydrogen system based on modified NSGA-II and CRITIC-TOPSIS. Energy Convers Manage 215:112892
87. Durmaz E, Akan Ş, Bakir M (2020) Service quality and financial performance analysis in low-cost airlines: an integrated multi-criteria quadrant application. Int J Econ Bus Res 20:168–191
88. Marjanovic I, Popovic Ž (2020) MCDM approach for assessment of financial performance of serbian banks. Bus Perform Financ InstS Eur. Springer.
89. Adali EA, Tus A (2021) Hospital site selection with distance-based multi-criteria decision-making methods. Int J Healthc Manag 14:534-544
90. Atanassov KT (1999) Intuitionistic fuzzy sets. In: Intuitionistic fuzzy sets (pp 1–137). Physica, Heidelberg
91. Yager RR (2013) Pythagorean membership grades in multicriteria decision making. IEEE Trans Fuzzy Sys 22(4):958–965. https://doi.org/10.1109/TFUZZ.2013.2278989
92. Kamali Saraji M, Streimikiene D, Kyriakopoulos GL (2021) Fermatean fuzzy CRITIC-COPRAS method for evaluating the challenges to industry 4.0 adoption for a sustainable digital transformation. Sustainability 13:9577.
93. Senapati T, Yager RR (2019) Fermatean fuzzy sets. J Ambient Intell Humaniz Comput 11:663–674
94. Saraji MK, Streimikiene D, Lauzadyte-Tutliene A (2021) A novel pythogorean Fuzzy-SWARA-CRITIC-COPRAS method for evaluating the barriers to developing business model innovation for sustainability. In: Handbook of research on novel practices and current successes in achieving the sustainable development goals. IGI Global.
95. Kamali Saraji M, Streimikiene D, Ciegis R (2022) A novel pythagorean fuzzy-SWARA-TOPSIS framework for evaluating the EU progress towards sustainable energy development. Environ Monit Assess 194:1-19
96. Keshavarz-Ghorabaee M, Amiri M, Hashemi-Tabatabaei M, Zavadskas EK, Kaklauskas A (2020) A new decision-making approach based on fermatean fuzzy sets and WASPAS for green construction supplier evaluation. Mathematics 8:2202
97. Ding B, Ferras Hernandez X, Agell Jane N (2021) Combining lean and agile manufacturing competitive advantages through Industry 4.0 technologies: an integrative approach. Prod Plan & Control, 1–17.

The manufacturer's authorised representative in the EU is Springer Nature Customer Service Centre GmbH, Europaplatz 3, 69115 Heidelberg, Germany. If you have any concerns regarding our products, please contact ProductSafety@springernature.com

Printed and bound by CPI Group (UK) Ltd, Croydon, CR0 4YY
30/09/2026
02225245-0004